Math Intervention for 5th Grade
Teaching Foundational Skills
A Teacher Resource

By Rod Rammage

Yellow Pencil Mathematics ^{TM}

Table of Contents

Introduction

The question is, as teachers, are we going to continue to let students slip through our system without a solid mathematical foundation? I am talking about the ones who don't easily understand math, and who have poor basic fact knowledge. Although they lack these integral skills for mathematics success, the expectation is that we can, and should, be able to teach them rigorous curriculum. I have been teaching and working with children for twenty years. My teaching experiences span many grade levels including all of those from grades five through eight. For the last seven years, I have worked in Texas as a Math Intervention Teacher in both fifth and sixth grade. I have been highly successful in working with under performing students to attain passing scores on the State of Texas Assessment of Academic Readiness (STAAR). District data has shown that these students, all of whom have never achieved passing scores on the state assessment, have increased their understanding with over 95% showing improvement and over 60% achieving passing scores on the STAAR.

I have created this resource over the past several years while working with my students. I am thrilled to be able to share my success with you. As a teacher of low performing students, I have always wanted an additional classroom resource like this. This tool is wonderful for both classroom teachers and intervention teachers. It can be used to supplement any mathematics curriculum in many different ways.
- An Intervention Program for fifth or sixth grade math
- An entire school RTI (Response to Intervention) math program
- A classroom resource to reinforce math skills and to fill in deficits in mathematical knowledge
- A homeschool math resource

Foundational Cornerstones - nine different mathematical skills along with my methods (lessons) for teaching them.
1. Addition
2. Subtraction
3. Multiplication
4. Division
5. Place Value
6. Expanded Form and Expanded Notation
7. Number Lines
8. Rounding
9. Adding and Subtracting Simple Fractions with unlike Denominators

Each lesson contains worksheets and full size answer keys.

Three skill tests—Beginning Screening, Middle Screening, and End Screening
Eight Bell Ringers—Fact practice papers students pick up and work on when they enter my room
An example of an Intervention Report
My Personal instructional views

Do...or Do Not

Yoda—Jedi Master
The Empire Strides Back

Teach...or Teach Not

Rod—Intervention Master
Yellow Pencil Mathematics

I don't know when I became a teacher; it just happened.

When I was young we studied pure mathematics—computation. Word problems were few and far inbetween. Most people forgot what they learned because they did not use the knowledge or did not know how to apply math in daily life. Over the years, the educational mathematics community has come to realize students must apply mathematics as they learn—hence word problems and manipulatives are a must.

More recently another concept has become popular. Rigor. For the highest achievement, problems must be rigorous—word problems requiring multiple mathematical concepts within the same problem. Because we want students to learn all they can, teachers have a busy curriculum and are taught to stay in their lane—not to teach above or below their grade. This is great if the student being taught is mathematically at the level they should be. But, if a student is not understanding, making mistakes and the word problems contain multiple concepts—for example, a problem containing multiplication, division and subtraction—what do they need help with? Multiplication? Division? Addition? Reading?

The truth is, many students have gaps in their mathematical knowledge which prevents them from being successful in mathematics. I perceive fifth grade as a pivotal year. Students who are struggling are starting to question their own intelligence—I have students tell me they are dumb because they don't do well in math. This breaks my heart because I know it is not true. Mathematical knowledge is a journey. Because someone walks slow, has to look at a map often, stop and ask directions, or has to ask for a guide does not mean they cannot complete the journey.

2

My main focus is to teach students to understand our number system and how it can be manipulated to their benefit.

I have designed this program to isolate and target essential skills. Once a student is proficient in the skill, teachers can then add rigor to expand students' ability.

The goal of intervention is to teach and practice the foundational skills students missed, are lacking in, did not learn, were absent that day, or for what ever other reason they haven't learned them so the regular math teachers can do their job.

At the start of this program, I always test students skills using the Beginning Screen Test—included. I use these tests, beginning, middle and end of year to diagnose student need and measure student growth. The Beginning, Middle, and End Screen Tests are mirror tests. The questions are the same; only the numbers are different. The questions are by skill in sets of three.

I test students multiplication facts knowledge using the 64 Mixed Multiplication Facts bellringer on pg 105. I keep a record of their test scores and multiplication times. I continue to test multiplication monthly. I always give students ten minutes to complete the multiplication activity. If a student does not complete the paper in less than ten minutes or she makes more than two mistakes I record an X. My goal is for students time to be under three minutes with no mistakes. Do not be concerned if a student makes marks on her paper, uses repeated addition, or counts fingers.

A Few More Thoughts

When asking students questions, give them plenty of time before helping them with the answer. If I am working with a large group I will not leave a student thinking long enough to embarrass them. If I am working in a small group or one on one with a student I will give them lots of time to think before I give them an answer.

I never teach at the board more than five minutes at a time. Anything more is just blab, blab, blab.

Have students work the same problems over and over. Have them work the same worksheet over and over. Memorize it? Oh please do. To memorize the way to solve a problem is to memorize an algorithm. Algorithms are great. Algorithms always remain the same. Algorithms can be broken down into steps. Algorithms are how we solve math problems.

When your students get it, use them as teachers as much as possible.

The best math resource you can purchase is a yellow pencil.

Lessons

We can teach mathematics in many different ways. My goal, with this program, is to focus on one that always works to create a corner stone for students to build from.

Section (A) Addition

Although most students will already have a working knowledge of addition, you can easily improve on their understanding. I never stress students to memorize addition or subtraction facts. Memorization of a number line and how to manipulate it is much more conducive to true mathematical understanding. For practice use the 64 Addition Facts bellringer pg on 101. Encourage students to use the number line when adding. I always tell students I am including a calculator the first time I hand this paper out. I wait for someone to ask for the calculator which opens a conversation about our number system being a number line, how easily we can make one, and how to use a number line to add numbers without ever making a mistake. Example: on the first problem 7 + 9, have students place one finger on 7 and from 7 but not including 7 count up 9 places to 16. On the board:

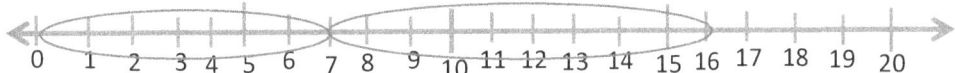

The answer is 16. Make sure students understand we are counting the spaces between the numbers. Point out to students how quick and easy this is. It is just like using a calculator I tell students, in my class, your brain is your calculator and I am going to teach you how to use it. Teach students to use the number line for addition and we will be able to build off of that learning for subtraction. Always, monitor students papers for mistakes. I have looked at so many of these I can spot a mistake in seconds. Repetition works for teachers too. A good way to start this is to pick out five or six problems to focus on as you walk around the room. If you don't monitor, many students will not put forth their best effort. Immediate feedback and correction is a must.

The worksheets, (A) 1 through 4, practice several different mathematical concepts: addition practice, place value practice, and penmanship. All are essential. I never let students slide with numbers I cannot easily read, are formed incorrectly, or that are leaking out of their place value column.

Always have students insert the decimal into numbers even when the number does not contain any decimal place values. Try this; put the number 309 on the board and ask students, where does the decimal go? Make sure you don't show any facial expression or display any body language. Don't be in a hurry, give them a minute. Wait, don't give them the answer. Write the number 00309 on the board. Again, ask them where the decimal goes. Now you begin to understand why one of my main mathematical focuses is Place Value. Without giving away the answer, ask someone to read the number for you. Three hundred nine. Don't accept three hundred and nine. On the board show them:

Explain to students, you said three what? Three Hundred. So, three has to be in the hundreds place, zero has to be in the tens place and nine must be in the ones place. The decimal is always after the ones place.

Hundreds	Tens	Ones	and
3	0	9	.

You might think they have this now but try the same exercise tomorrow and again next week. Don't get frustrated. Try to work in the question, where does the decimal go as often as you can. A solid understanding in Place Value and how numbers are manipulated within the system is essential to understanding mathematics. I never tell students, when adding or subtracting, to line up the decimals without explaining to them that what they are really lining up is the Place Value.

Monitor and instruct students as they practice adding numbers. Make sure students understand that each value column can only hold the single digit numbers (0—9) and how to carry over to the next place value. If you don't physically look at individual papers you don't know if students really understand. Watch out for students adding left to right instead of right to left.

With each worksheet, I have included a full page answer key for you (with math tracks as I would have worked the problems) to display on the board so students can check their work. Immediate feedback is essential to understand mathematics.

To make the statement, "oh this is easy", or "you should know this", or "you should have learned this by now" is the equivalent to calling someone stupid. How well do you learn from a person who calls you stupid? Students can only learn to love mathematics if you are patient and show understanding of their struggle. No mathematical concept is easy until you understand it and have learned to use it.

Section (B) Subtraction

Many students struggle with subtraction. Have students practice the bellringer 64 Subtraction Facts on pg 102. This worksheet is a mirror image of the 64 Addition Facts worksheet on pg 101. As with addition, teach students to use the number line as a calculator. For 16 – 9, have students put one finger on 9 and count up to 16. The answer is 7.

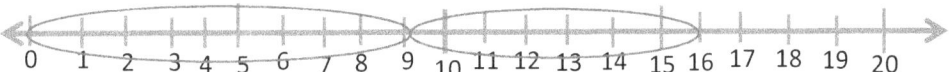

Demonstrate to students how quickly and easily this can be done. The objective is to teach students to be able to picture and understand how addition and subtraction are related and behave on a number line. Monitor students papers for mistakes.

Teach students to practice subtraction worksheet (B) 1 through 4. Continue to stress place value and penmanship. Monitor that students are regrouping correctly. Watch for students working problems left to right instead of right to left.

Don't worry, if you do a bad job, if you are not patient, or if you are not kind, **you** will not be remembered.

Section (C) Multiplication

Learning multiplication is the hardest. Learning multiplication is the key to the mathematical kingdom. Unlike addition and subtraction facts, students must memorize multiplication facts. The good news is, because of the Associative Property of Multiplication--a x b = b x a--students only need to memorize 36 facts–see bellringer 36 Facts in Order 2s Through 9s on pg 103.

For 0, zero is always the hero in both multiplication and division. Your students already know their 1s and 10s. You can teach them all their 2 X 11 through 9 X 11 in one minute. Once they know their 11s, teach them their 12s by adding one multiple to the 11. Example: 3 X 12 = 3 X 11 + 3.

Show students how multiplication works on a number line.

6 X 3 = 3 + 3 + 3 + 3 + 3 + 3 = 18

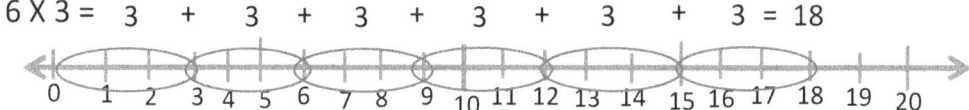

Now, explain that doing this to solve problems would become too strenuous. I spend the first week or so of my class every year using multiplication flash cards. I take all cards out of the deck except the 2s through 9s. Then, using the associative property, I create two decks out of each one. I let students work in pairs. Students **must** be taught how to quiz other students. I sit down to teach one student and have others gather around me. I start flashing multiplication facts. When they get the fact right, I put that fact in the-–know—pile. When the student misses a fact I put it in the-–don't know—pile. After I accumulate three cards—facts—in the don't know pile, I stop, pick up that pile and quiz those three facts until the student has them down. Then, I mix those three facts back into the, know, deck and continue to quiz the facts. To start, I will let the students who are having the most trouble be the quizzers. At first, most students will love working with flash cards. I continue this activity as long as students retain interest. You must monitor this activity. Sometimes, I can get a week or even two out of this activity. When students start to lose interest and goof off, it is time to move on.

Students should complete the, Multiple Practice bellringer on pg 106, as often as possible. To teach this, I place a blank copy of the bellringer on my document camera and show students that multiples are repeated addition. Then, I tell students I am going to race them, but that I will give them a head start. I walk around the room as students work offering words of encouragement: wow, look at you go, doing good, watchout I can't read that four, fantastic, I'm going to start in two minutes, you just might beat me. I return to my document and quickly complete the multiples as students check their work. I continue to encourage students as I work. Don't be afraid to congratulate the students who beat you. I **never** let students complete this worksheet without putting a correct copy in front of them—either on the board or I hand them a completed copy. After students complete this activity, I write the numbers 1—10 across the top and point out I can easily create my own multiplication chart. You might notice this bellringer is in the same format as the multiplication chart I included with this program. Don't be afraid to let students, who were not able to complete this activity, use a copy of the multiplication chart. I repeat this activity every chance I get. Not only will this activity help students learn multiplication, this knowledge will be invaluable when you teach division or adding and subtracting fractions.

Almost every day, I do a five minute count down time test for the bellringer, 36 Multiplication Facts in Order 2s Through 9s on pg 103. I always have a copy of the bellringer, 36 Mixed Multiplication Facts 2s Through 9s on pg 104, on the back of this paper. I teach students, once they get the double, all they have to do is count to get the rest. Don't be afraid to let them use their fingers. This is an activity; it is not really a test. Don't be afraid to walk around and tell students the answer to a double if they don't know it. When a student calls out done, I call out the time. As soon as a student finishes I quickly scan the paper and circle any mistakes and ask them to correct them. At this point, if a student is still struggling I hand them a multiplication chart. As soon as a student has a completed, correct copy, I ask them to complete the 36 Mixed on pg 104 on the back. Students are free to use the copy of the 36 in Order to complete the 36 Mixed. Every fact on the 36 Mixed is on the 36 in Order.

I do not teach Place Value when teaching two digit by three digit multiplication worksheets (C) 1—5. Place Value will confuse students when they start multiplying with decimals. I stress the importance of penmanship and keeping digits in columns. The biggest mistakes I see students make, other than addition and multiplication errors, is forgetting to put in zeros as place holders and not multiplying the correct numbers all the way through. See the following illustrations.

326 X 42 =

```
326        326
x 2        X 40
652   +    13,040
```

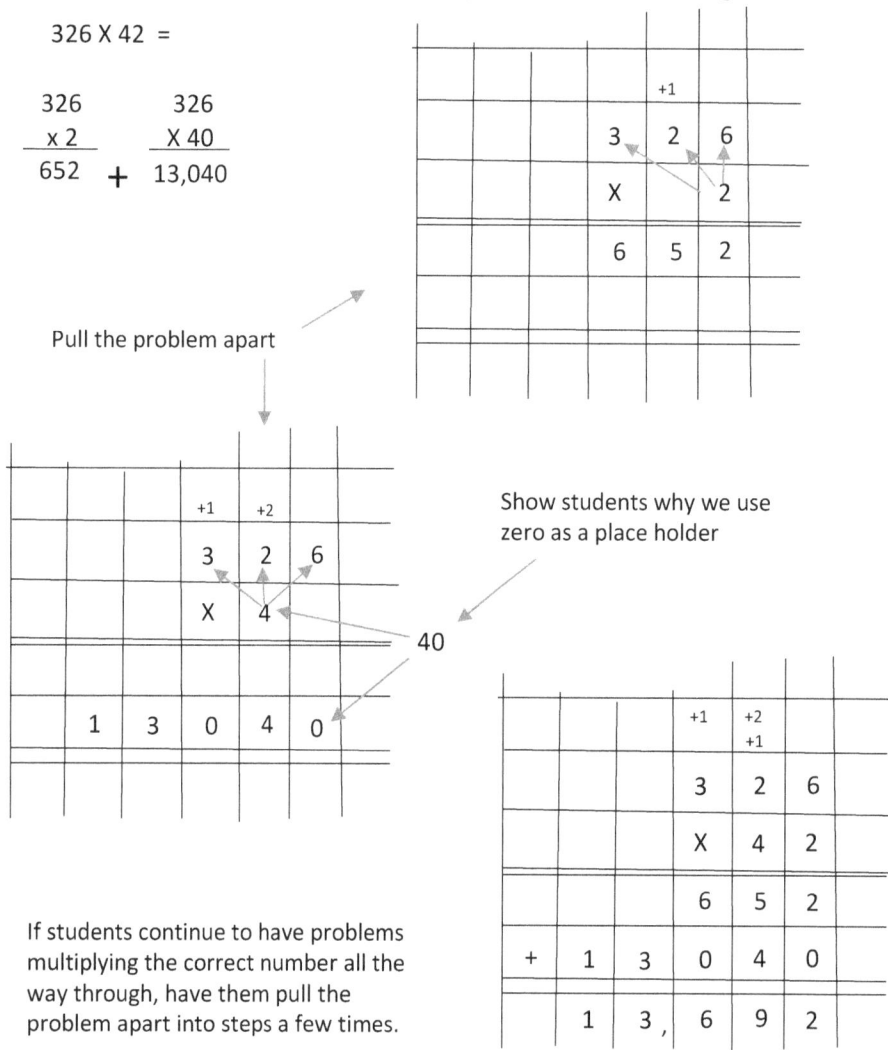

Pull the problem apart

Show students why we use zero as a place holder

40

If students continue to have problems multiplying the correct number all the way through, have them pull the problem apart into steps a few times.

Reteach Reteach Reteach Reteach Reteach—Don't get frustrated—Reteach When students continue to struggle, I have them repeat worksheet C1 over and over and over. Don't be afraid to let students use a multiplication chart when practicing these worksheets if they need it.

10

Your importance to an individual is dependent upon your ability to convince that person of their importance to you.

Section (D) Division

Once students have some success with multiplication facts, start letting them use the Division Starter bellringer pg on 108. You should note, the 64 Mixed Multiplication Facts 2s Through 9s, the Division Starter, and the Division Practice bellringer are the same format.
(7 X 3 = 21, 7 X 3 = 21, and 21 / 3 = 7
When students first start practicing this bellringer, teach them how to use the multiplication chart to get their answers. After students get going show them how much faster—if they know the fact—it is to say, what times three equals twenty-one and get the answer than it is to use the multiplication chart. Students only need to practice this bellringer a couple of times.

Students are finally ready for the Division Practice bellringer on pg 109. To introduce this bellringer display a division sign on the board. Show students how much a division symbol and a fraction look alike.

$$\frac{\bullet}{\bullet} \qquad \text{Looks like} \qquad \frac{1}{2}$$

Tell students, that's right. The division symbol looks like a fraction. A fraction is a division problem.

$$\frac{21}{3}$$

Twenty- one divided by three can be shown as

What times three equals twenty-one? 7. So we know, twenty-one divided by three equals seven.   Practice   Practice   Practice

The long division algorithm is a strenuous, complicated and confusing undertaking. On the following pages, I show you how to take the strain out of the algorithm so students can divide with confidence, ease, accuracy and speed.

Division with Factors and Grouping

The Quotient and Dividend live in the same Place Value. The Divisor has a Place Value of its own.

Always use numbers in Place Value. Never use Xs as place holders. If a Place Value is blank, 0 is the correct place holder.

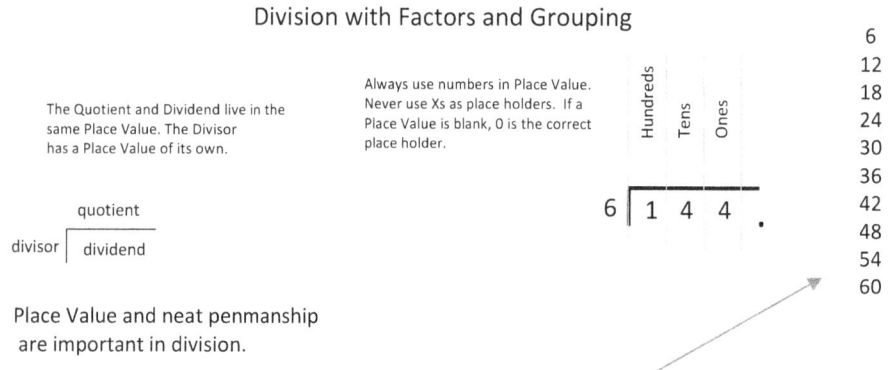

quotient

divisor | dividend

Place Value and neat penmanship are important in division.

List the first 10 factors of the divisor. Students can use repeated addition if they don't know them.

Think about the number 1 in the hundreds place.
Can you take 6 out of a group of 1?

No, then put a 0 in the hundreds place.

So, the 4 from the tens place can come help by making a group of 14.

Can you take 6 out of a group of 14?

Yes. How many times?

Two 6s equals 12 so subtract 12.

Fourteen minus twelve equals two.

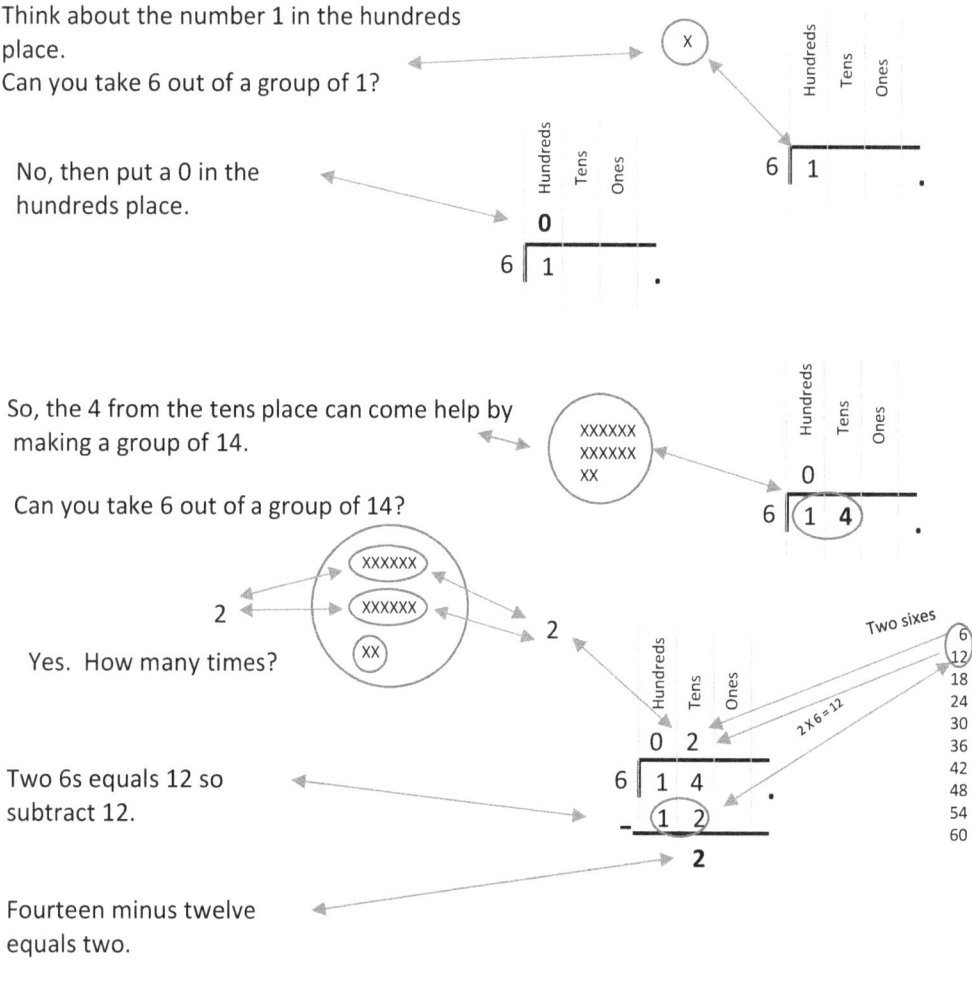

Division With Factors and Grouping

Can you take 6 out of 2?   No.

So, the 4 from the 1s place can come down and help by creating a group of 24.

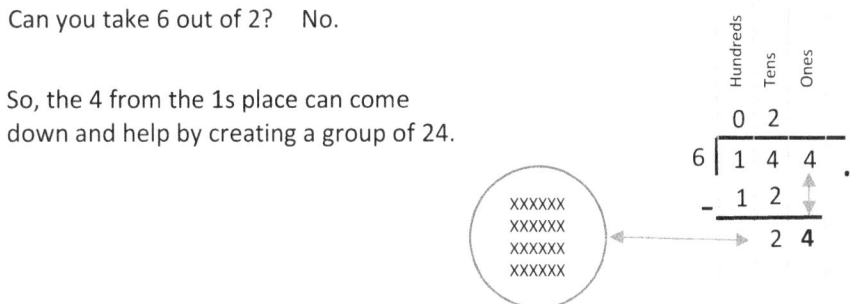

Can you take 6 out of 24?

How many times?   4

Four sixes equal 24, so subtract 24.

24 − 24 = 0

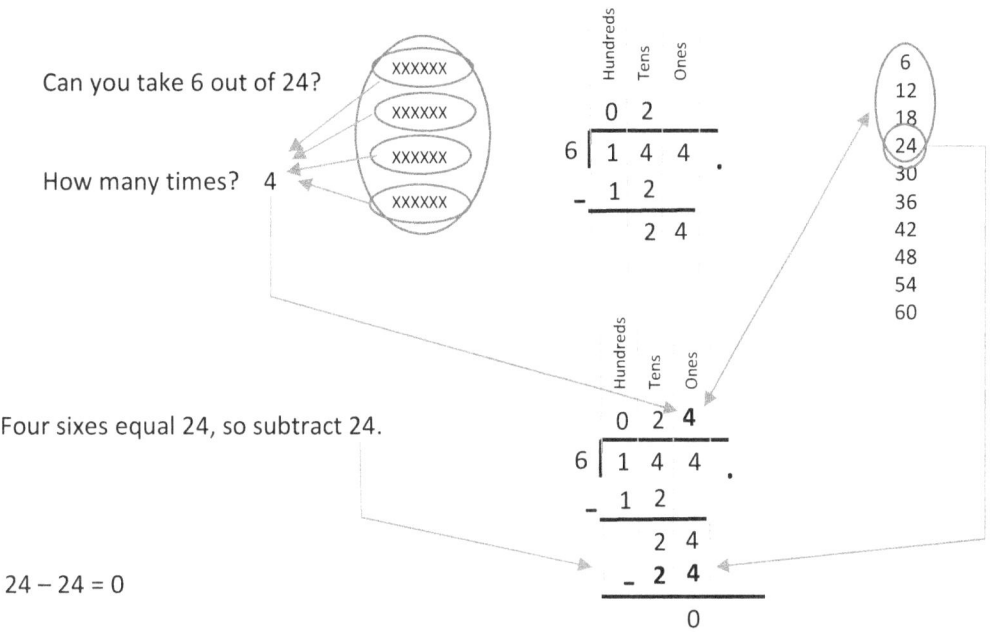

You have no number left to bring down. You are done and your answer is 24.

Once you have taught showing students grouping, teach students to only use the list to find their answers. (Do **NOT** have students draw Xs and circle groups. Teach students to use the list of factors to find their answers.)

Students are now ready to start practicing worksheets (D) 1—5.
The hardest part of this method is for students to create an
accurate multiple list of the divisor. Most of the time, unless
students are sitting and working with me, I let them use a multiplication
chart to check or list their divisor. I wrote the following
dialog as it would sound if I were helping a student.

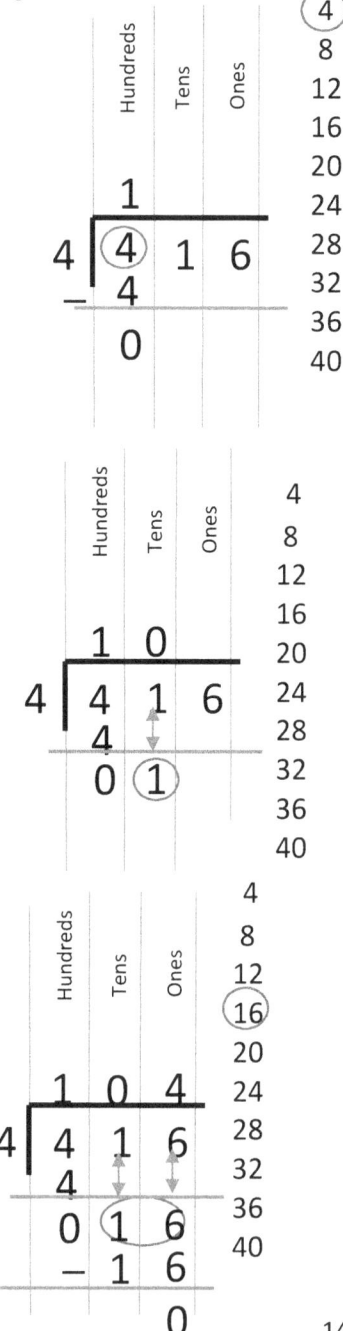

How big is the first group we need
to look at?   4.
Look at the list. Can I take 4 out of
 that group of 4. Yes. Can I take
8 out of that group of 4? No
Go back, circle and then count.
One. So, put one up and subtract 4.
We now have 0 in our group.

Bring the 1 down. We now have
1 in our group. Look at the list. Can
We take 4 out of a group of 1. No.
What is the number for no number.
Zero. So, put a 0 up and you can
bring the 6 down.

We now have a group of 16. Look at
the list. Can you take 4 out of 16?
Yes. 8? Yes. 12? Yes. 16? Yes. Can
you take 20 out of a group of 16. No.
Then go back circle 16 and count. We
have four 4s in 16 so put 4 up and
subtract 16. That equals 0 and we
have nothing left to bring down.

Practice D 1—4 over and over
and over and over and over

Save D5 for when students
are ready for a challenge.

Mathematics is a journey. Be a good guide.

Section (E) 1—2 Place Value

With the Place Value worksheets, have students use the decimal as their anchor point. Find the decimal in a number and work out in both directions. Don't be surprised if students still struggle with where to put the decimal in numbers when no decimal is shown.
Practice   Practice   Practice   Practice

Section (F) 1—4 Expanded Form and Expanded Notation

By using Place Value charts students are able to see how each number is related to its place value as they decompose it. When you have student study Expanded Form first, Expanded Notation is an easy next step and deepens students understanding of the actual value of numbers. Use the Expanded Notation charts to start a discussion about how numbers move from one place value to another. Point out that we have a base ten number system. From right to left ones times ten equal tens, tens times ten equal hundreds, hundreds times ten equal thousands, and so on. From left to right thousands divided by ten equal hundreds, hundreds divided by ten equal tens, tens divided by ten equal ones, and so on.

Section (G) 1—5 Number Lines

Our number system is a base 10 number line. Because one of my main objectives with this program is to teach students to understand this system I am limiting this teaching to lines with increments of 10.

| 0 | 1 | 2 | 3 | 4 | 5 | 6 | 7 | 8 | 9 | 10 |

This is a good time to start learning to divide by 10s by dividing out zeros.

$$10 = 1$$

This line is in ten increments. If we divide by ten we get one, so we should number the increments by 1s.

For lessons G1—G3, students will have trouble labeling increments unless you teach them how.

$$100 = 10$$

This line is in ten increments. If we divide by ten we get ten, so we should number the increments by 10s.

$$200 = 20$$

This line is in ten increments. If we divide by ten we get twenty, so we should number the increments by 20s.

$$500 = 50$$

This line is in ten increments. If we divide by ten we get fifty, so we should number the increments by 50s.

$$1000 = 100$$

This line is in ten increments. If we divide by ten we get one hundred, so we should number the increments by 100s.

For lesson G4 and G5 my main objective is to demonstrate how small a difference decimals make in the movement of a point on a line. After the first couple of problems, don't label the entire line. Label the ten increments between the two closest whole numbers. Only label to the tenths place. Discuss how little difference is made by the hundredths and thousandths place.

Look at that. You have them all set to start estimation. Use worksheets GH 1 and GH 2 to practice estimation.

Teaching is more than a passing of knowledge, it is a passing of Humanism.

Section (H) Rounding

The goal of teaching students to round numbers is to enable students to estimate and quickly solve math problems with little effort. For the most part, in life, we estimate to eliminate the use of decimal numbers.

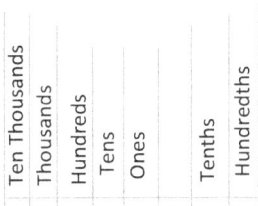

A basic knowledge of place value is imperative for accurate estimation.

When, rounding, I teach students to use this scale.

$$3\ 2\ \overset{+1}{\underline{4}}\ .\ \textcircled{8}\ 1 \quad = \quad 325$$

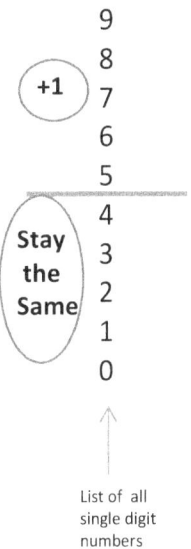

To round 324.81 to the ones place, I would:
First, underline the four in the ones place
Second, strike out the numbers being eliminated
Third, circle the number which controls the process
Fourth, add one to the ones place or leave it the same
Fifth, write down my answer

List of all single digit numbers

Before teaching students to use the scale for rounding, we **must teach** them why it works. See the next page.

To teach students why the scale for rounding works, use the following two examples.

Round the number 18.82 to the nearest whole number.

What we are **really asking**: is the number 18.82 closer to the whole number 18 or the whole number 19?

18.82 is closer to 19 so the answer is 19

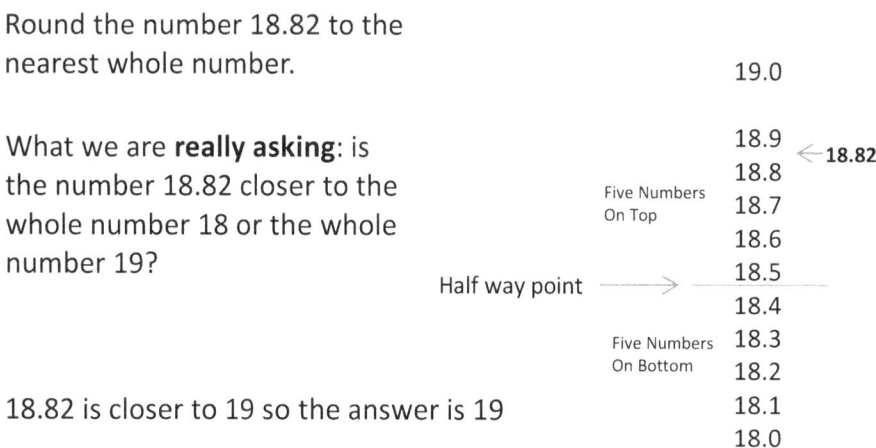

This concept can also be shown on a number line.

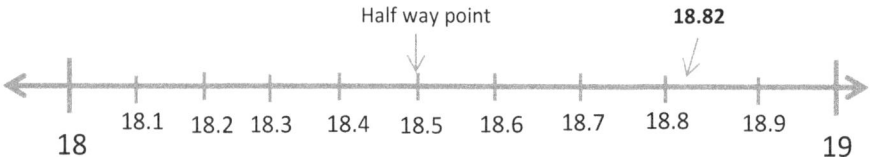

18.82 is closer to 19 so the answer is 19

Use worksheets H1—H3 to teach rounding. Use worksheets H4 and H5 to teach estimation. The first time I give students worksheet H4 I do not give them any verbal instructions other than please complete this worksheet. Most students will line up their decimals and work the problems. The next day, I hand out the same paper and tell students they made a mistake. What do the directions mean. Estimate. We redo the paper using estimation to make it easier and faster. Students really start to understand the power of estimation in worksheet H5.

To have a student in your room is to own their time—an actual piece of their life. Don't think your time is anymore valuable than theirs.

Section (I) Adding and Subtracting Fractions

At first, I always use Least Common Multiple, (LCM), to teach students to add and subtract fractions with unlike denominators.
This method **always** works and students usually don't have to simplify their answer. Pryor to teaching this lesson, you have been having students practice the Multiple Practice bellringer on pg 106. Now use worksheet I1, Least Common Multiple, to teach students how to find LCM.
Students are now ready to learn to add and subtract fractions. I will try to give you an example of how I would teach this concept.

LCM	4	4, 8, 12
	3	3, 6, 9, 12

$$\frac{1}{4} + \frac{2}{3}$$

Have students find the LCM of the denominators. Now tell students, next to the fraction put a fraction bar plus a fraction bar equals a fraction bar.

$$\underline{\quad} + \underline{\quad} = \underline{\quad}$$

Now, put in your LCM as your denominator for all three. This step prevents students from adding their denominators.

$$\frac{\quad}{12} + \frac{\quad}{12} = \frac{\quad}{12}$$

Now, we need to create equivalent fractions using the new denominator

$$\frac{1}{4} = \frac{\quad}{12} \quad \text{and} \quad \frac{2}{3} = \frac{\quad}{12}$$

$$\frac{1^{\times 3}}{4_{\times 3}} = \frac{}{12} \quad \text{and} \quad \frac{2^{\times 4}}{3_{\times 4}} = \frac{}{12}$$

What times four equals twelve? Three. If we multiply the denominator by three then we must multiply the numerator by three. One times three is three so our new equivalent fraction is three twelfths. Repeat this process for the other fraction.

$$\frac{1}{4} = \frac{3}{12} \quad \text{and} \quad \frac{2}{3} = \frac{8}{12}$$

We are now ready to add the numerators of our new equivalent fractions.

$$\frac{3}{12} + \frac{8}{12} = \frac{11}{12}$$

Completed example:

LCM	$\frac{4}{3}$	4, 8, 12 3, 6, 9, 12		12
$\frac{1^{\times 3}}{4_{\times 3}} + \frac{2^{\times 4}}{3_{\times 4}}$		$\frac{3}{12} + \frac{8}{12} = \frac{11}{12}$		$\frac{11}{12}$

Do not be in a hurry to move students to simplifying or working more difficult problems. I have students work the same three papers over and over and over. After students have had plenty of guided practice, worksheet i4 makes a great bellringer. What if they memorize it? I would love that as long as they can show all of the work. Do not let students become lazy and stop putting in the little times. I always keep extra copies. I have even had students ask if they could take extra copies home to show their parents what they can do.

Addition

Name_____

1) 8,054 + 986 =

Hundred Thousands	Ten Thousands	Thousands	Hundreds	Tens	Ones	And	Tenths	Hundredths	Thousandths

2) 8,745.3 + 89.563 =

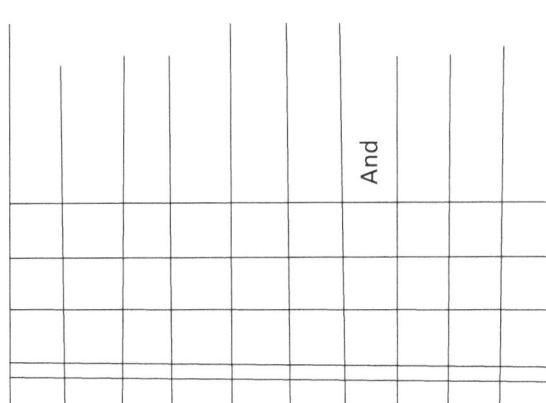

3) 45.432 + 896 =

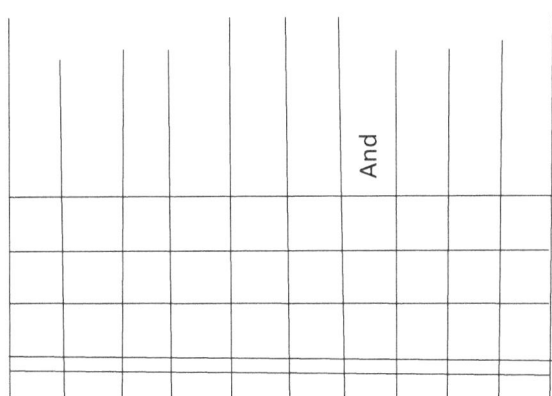

Addition

A 1

Name ___Key___

1) 8,054 + 986 =

Hundred Thousands	Ten Thousands	Thousands	Hundreds	Tens	Ones	And	Tenths	Hundredths	Thousandths
		¹8	¹0	¹5	4				
		+	9	8	6				
		9,	0	4	0				

2) 8,745.3 + 89.563 =

		Thousands	Hundreds	Tens	Ones	And	Tenths	Hundredths	Thousandths
		8	¹7	¹4	5	.	3	0	0
			+	8	9	.	5	6	3
		8,	8	3	4	.	8	6	3

3) 45.432 + 896 =

		Hundreds	Tens	Ones	And	Tenths	Hundredths	Thousandths
			¹4	5	.	4	3	2
		+	¹8	9	6	.	0	0	0
			9	4	1	.	4	3	2

Addition

Name _____

1) 786 + 56.904 =

Hundred Thousands	Ten Thousands	Thousands	Hundreds	Tens	Ones	And	Tenths	Hundredths	Thousandths

2) 7834.964 + 904 =

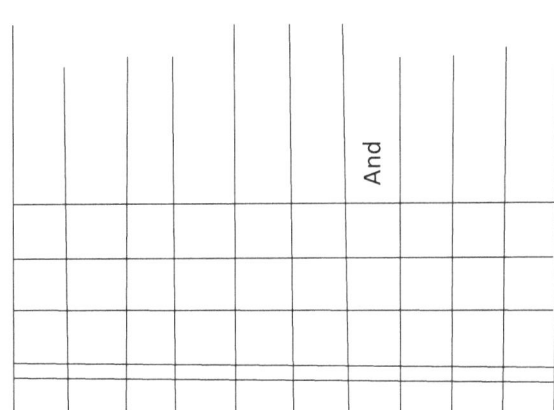

3) 90.899 + 0.9 =

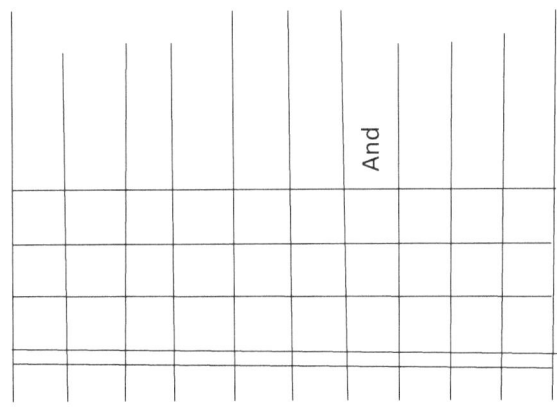

Addition

A 2

Name _____

1) 786 + 56.904 =

Hundred Thousands	Ten Thousands	Thousands	Hundreds	Tens	Ones	And	Tenths	Hundredths	Thousandths
			7	8	6	.	0	0	0
			+	5	6	.	9	0	4
			8	4	2	.	9	0	4

2) 7834.964 + 904 =

		Thousands	Hundreds	Tens	Ones	And	Tenths	Hundredths	Thousandths
		7	8	3	4	.	9	6	4
		+	9	0	4	.	0	0	0
		8,	7	3	8	.	9	6	4

3) 90.899 + 0.9 =

			Tens	Ones	And	Tenths	Hundredths	Thousandths
			9	0	.	8	9	9
			+	0	.	9	0	0
			9	1	.	7	9	9

Addition

A 3

Name _____

1) 90.7 + 345.005 =

Thousands

Tens   Thousandths

Hundred Thousands

Tenths   Ones

Ten Thousands

Hundredths

Hundreds

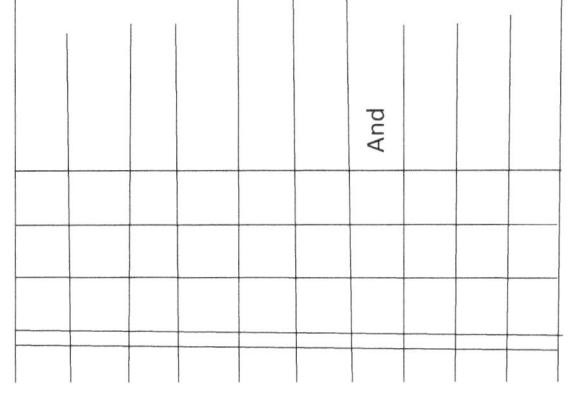

2) 999.998 + 0.009 =

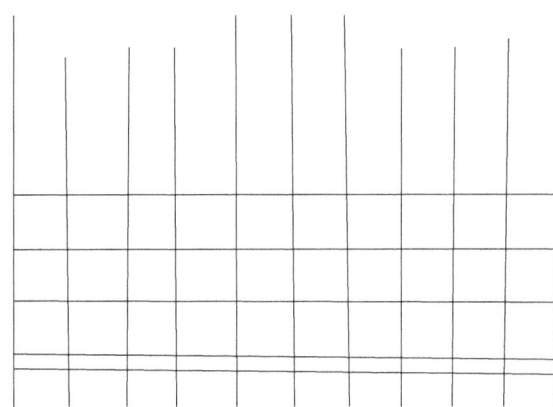

3) 430.6 + 8,823.97 =

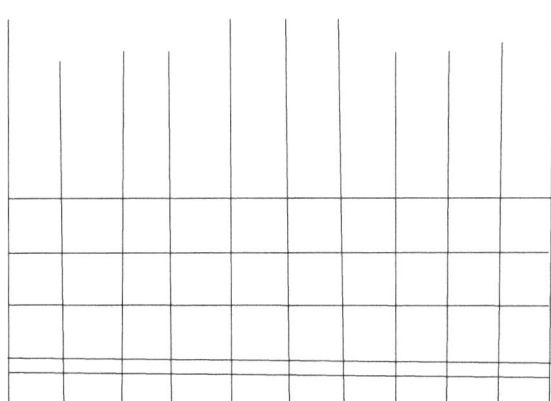

Addition

A 3

Name _____

1) 90.7 + 345.005 =

Hundred Thousands	Ten Thousands	Thousands	Hundreds	Tens	Ones	And	Tenths	Hundredths	Thousandths
						.			
				9	0	.	7	0	0
		+	¹3	4	5	.	0	0	5
			4	3	5	.	7	0	5

2) 999.998 + 0.009 =

		Thousands	Hundreds	Tens	Ones	And	Tenths	Hundredths	Thousandths
						.			
			¹9	¹9	¹9	.	¹9	¹9	8
				+	0	.	0	0	9
		1,	0	0	0	.	0	0	7

3) 430.6 + 8,823.97 =

		Thousands	Hundreds	Tens	Ones	And	Tenths	Hundredths
						.		
			4	3	¹0	.	6	0
		+	¹8	8	2	3	.	9	7
		9,	2	5	4	.	5	7

Addition

Name _____

1) 9,862 + 345.64 =

Thousands

Tens   Thousandths

Hundred Thousands

              Ones
Tenths

        Ten Thousands

Hundredths

        Hundreds

						And			

2)   472.8 + 215.908 =

3)   78,070.4 + 96.5 =

4)   99 + 999.99 =

5)   9,048.76 + 215 =

Addition

A 4

Name <u>**Key**</u>

1) 9,862 + 345.64 =

Hundred Thousands	Ten Thousands	Thousands	Hundreds	Tens	Ones	And	Tenths	Hundredths	Thousandths
		$\overset{1}{9}$	$\overset{1}{8}$	6	2	.	0	0	
		+	3	4	5	.	6	4	
	1	0,	2	0	7	.	6	4	

2) 472.8 + 215.908 =

$$\overset{1}{}472.800$$
$$+\,215.908$$
$$\overline{688.708}$$

3) 78,070.4 + 96.5 =

$$\overset{1}{}78070.4$$
$$+96.5$$
$$\overline{78,166.9}$$

4) 99 + 999.99 =

$$\overset{1}{}\overset{1}{}99.00$$
$$+999.99$$
$$\overline{1,098.99}$$

5) 9,048.76 + 215 =

$$\overset{1}{}9048.76$$
$$+215.00$$
$$\overline{9,263.76}$$

Subtraction

Name _____

1) 8,054 - 986 =

Hundred Thousands	Ten Thousands	Thousands	Hundreds	Tens	Ones	And	Tenths	Hundredths	Thousandths

2) 95.3 - 89.563 =

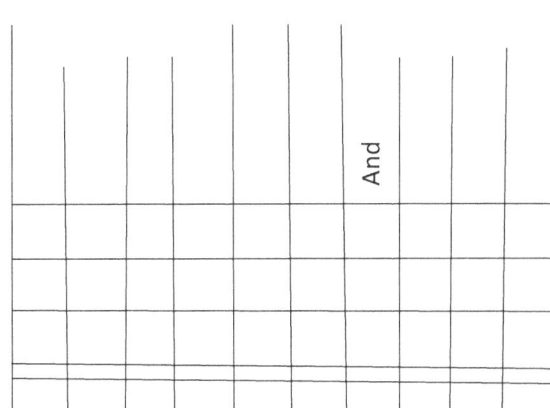

3) 1,145.432 - 896 =

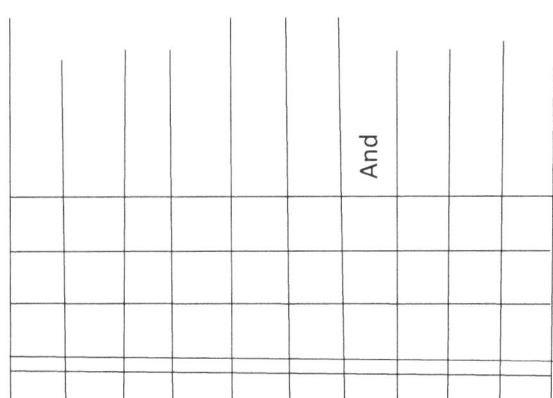

Key

Name _____

1) 8,054 - 986 =

Hundred Thousands	Ten Thousands	Thousands	Hundreds	Tens	Ones	And	Tenths	Hundredths	Thousandths
		⁷8	⁹¹0	¹⁴5	¹4				
		−	9	8	6				
		7,	0	6	8				

2) 95.3 - 89.563 =

			Tens	Ones	And	Tenths	Hundredths	Thousandths
			⁸9	¹⁴5	.	¹²3	⁹¹0	¹0
			− 8	9	.	5	6	3
				5	.	7	3	7

3) 1,145.432 - 896 =

	Thousands	Hundreds	Tens	Ones	And	Tenths	Hundredths	Thousandths
	⁰1	¹⁰1	¹³4	¹5	.	4	3	2
		8	9	6	.	0	0	0
		2	4	9	.	4	3	2

Subtraction

B 2

Name _____

1) 80.54 - 9.86 =

Hundred Thousands	Ten Thousands	Thousands	Hundreds	Tens	Ones	And	Tenths	Hundredths	Thousandths

2) 95 - 89.563 =

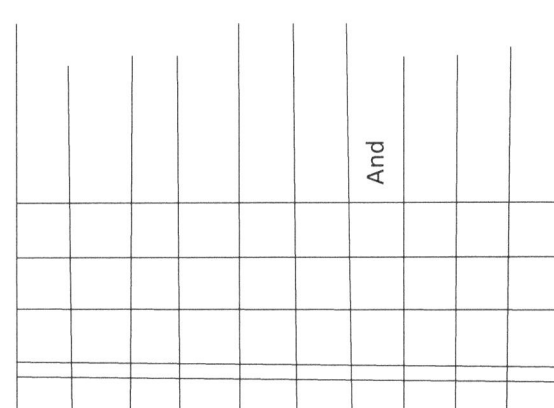

3) 1140 - 0.9 =

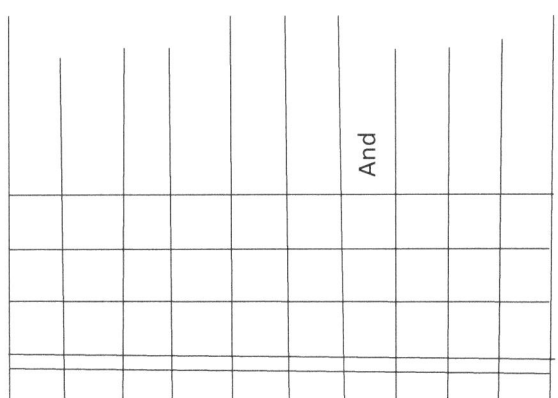

B 2

Name

1) 80.54 - 9.86 =

Hundred Thousands	Ten Thousands	Thousands	Hundreds	Tens	Ones	And	Tenths	Hundredths	Thousandths
				7̷8	¹0⁹	.	5¹⁴	¹4	
				−	9	.	8	6	
				7	0	.	6	8	

2) 95 - 89.563 =

				Tens	Ones	And	Tenths	Hundredths	Thousandths
				9⁸	5¹⁴	.	¹0⁹	¹0⁹	¹0
				− 8	9	.	5	6	3
				5		.	4	3	7

3)   1,140 - 0.9 =

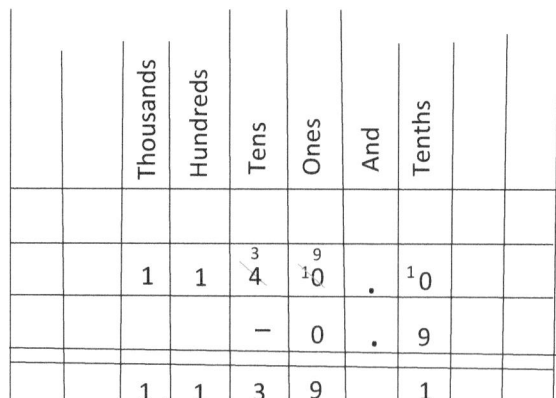

	Thousands	Hundreds	Tens	Ones	And	Tenths	
	1	1	4³	¹0⁹	.	¹0	
			−	0	.	9	
	1,	1	3	9	.	1	

Subtraction

B 3   Name _____

1) 6,890 - 799.07 =

Thousands   Ten Thousands

Ones   Thousandths

Tenths   Hundreds

Hundred Thousands

Tens   Hundredths

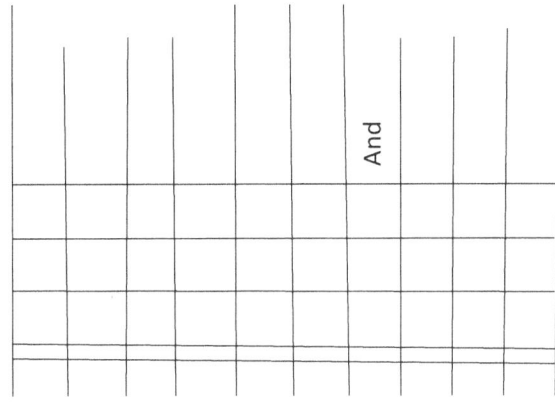

2) 769 - 1.604 =

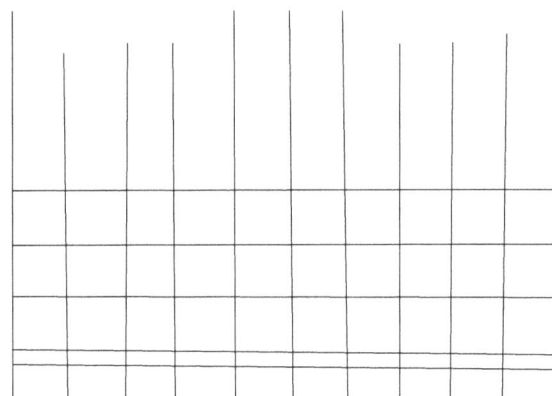

3) 78,432.4 - 9,311 =

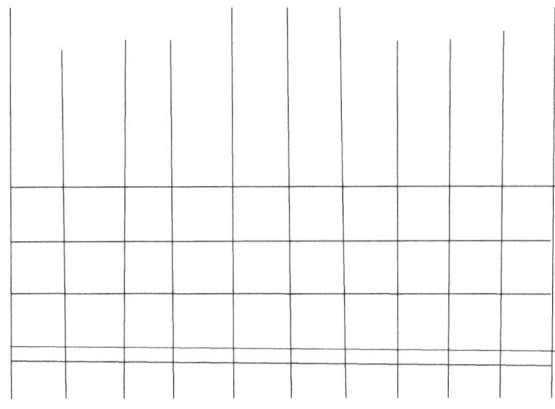

Subtraction

Name _____ **Key** _____

1) 6,890 - 799.07 =

Hundred Thousands	Ten Thousands	Thousands	Hundreds	Tens	Ones	And	Tenths	Hundredths	Thousandths
		6	^{7}8	^{18}9	$^9{}^10$	.	$^9{}^10$	^{10}	
		−	7	9	9	.	0	7	
		6 ,	0	9	0	.	9	3	

2) 769 - 1.604 =

			Hundreds	Tens	Ones	And	Tenths	Hundredths	Thousandths
			7	6	^{8}9	.	$^9{}^10$	$^9{}^10$	^{10}
				−	1	.	6	0	4
			7	6	7	.	3	9	6

3) 78,432.4 - 9,311 =

	Ten Thousands	Thousands	Hundreds	Tens	Ones	And	Tenths		
	^{6}7	^{1}8	4	3	2	.	4		
	−	9	3	1	1	.	0		
	6	9 ,	1	2	1	.	4		

Subtraction

B 4   Name _____

1) 1,111.11 -   799.07 =

Ten Thousands

Thousands

Ones   Thousandths

Tenths   Hundreds

Hundred Thousands

Tens   Hundredths

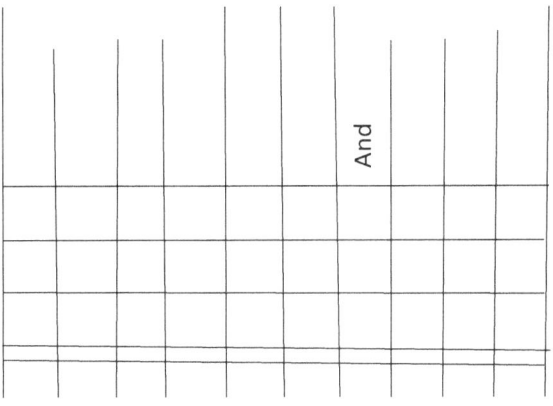

2)   384.22 - 23.9 =   3)   22 - 0.99 =

4)   748.543 - 725.9 =   5)   78.96 - 29 =

Subtraction

B 4

Name _____

1) 1,111.11 - 799.07 =

Hundred Thousands	Ten Thousands	Thousands	Hundreds	Tens	Ones	And	Tenths	Hundredths	Thousandths
		^{0}1	^{10}1	^{10}1	^{1}1	.	^{0}1	^{1}1	
		−	7	9	9	.	0	7	
			3	1	2	.	0	4	

2) 384.22 - 23.9 =

```
       3 1
3 8 4 . 2 2
- 2 3 . 9 0
3 6 0 . 3 2
```

3) 22 - 0.99 =

```
   1 9
   1   1
2 2 . 0 0
- 0 . 9 9
2 1 . 0 1
```

4) 748.543 - 725.9 =

```
     7 1
7 4 8 . 5 4 3
- 7 2 5 . 9 0 0
  2 2 . 6 4 3
```

5) 78.96 - 29 =

```
   6
   7 1
7 8 . 9 6
- 2 9 . 0 0
4 9 . 9 6
```

Multiplication

Name _____

1) 325 X 42 =

```
  325        325
  x 2        X 40
_____    _____
      +
```

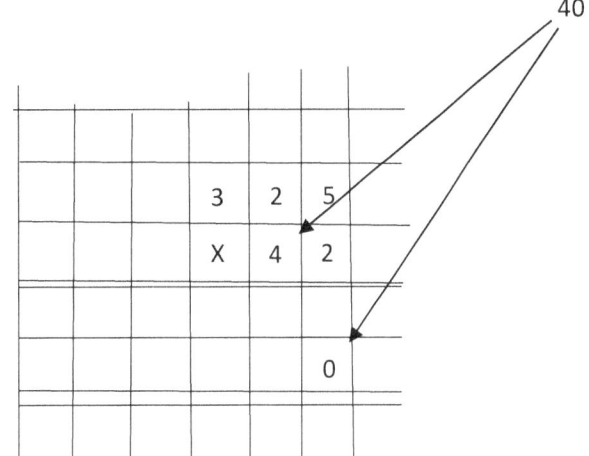

40

```
  3 | 2 | 5
  X | 4 | 2

          0
```

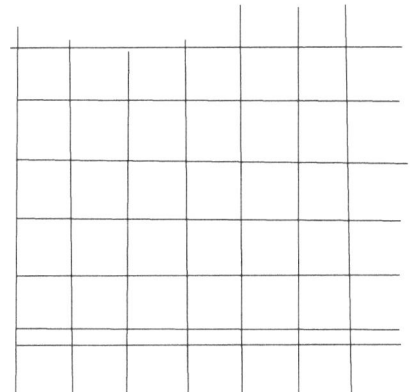

2) 785 X 28 =

3) 324 X 46 =

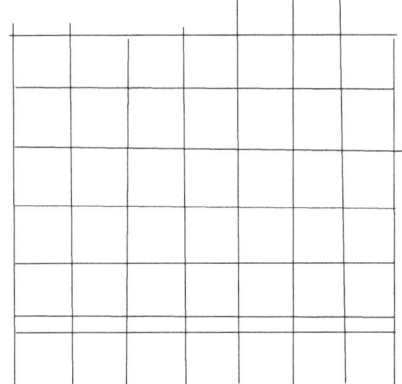

C 1   Name _____

1) 325 X 42 =

$$\begin{array}{cc} 325 & 325 \\ \underline{x\,2} & \underline{X\,40} \end{array}$$

+

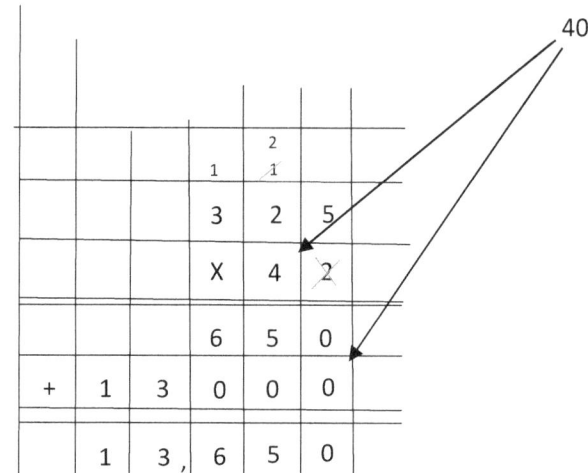

40

			$_1$	$^2_{\cancel{1}}$	
			3	2	5
			X	4	$\cancel{2}$
			6	5	0
+	1	3	0	0	0
	1	3 ,	6	5	0

		$^1_{\cancel{6}}$	$^1_{\cancel{4}}$	
		7	8	5
		X	2	$\cancel{8}$
$_1$	6	2	8	0
+	1	5	7	0	0
	2	1 ,	9	8	0

2) 785 X 28 =

3) 324 X 46 =

		$\cancel{1}$	$^1_{\cancel{2}}$	
		3	2	4
		X	4	$\cancel{6}$
$^1_1$	$^1_9$	4	4	
+	1	2	9	6	0
	1	4 ,	9	0	4

Multiplication

C 2   Name _____

1) 978 X 36 =

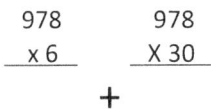

$$\begin{array}{r} 978 \\ \times\ 6 \\ \hline \end{array} \qquad \begin{array}{r} 978 \\ X\ 30 \\ \hline \end{array}$$

+

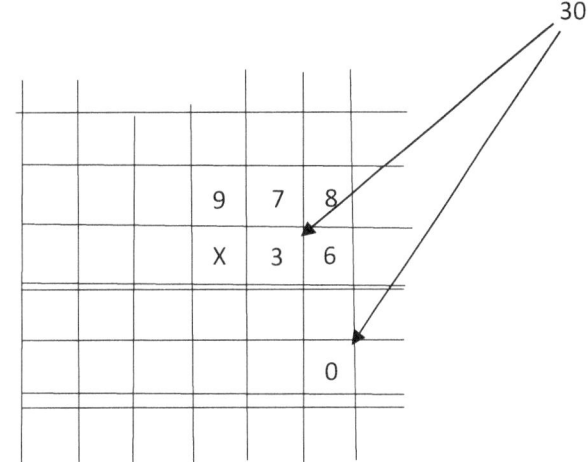

2) 129 X 94 =

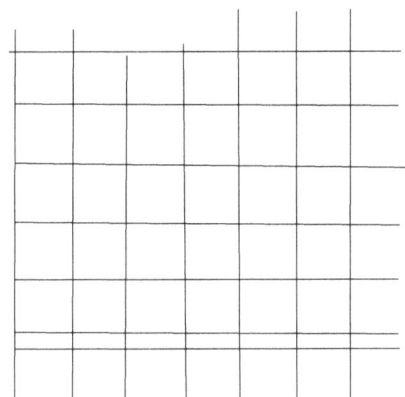

3) 456 X 78 =

Yellow Pencil Mathematics

Multiplication

Name_____ Key

1) 978 X 36 =

978   978
x 6   X 30
      +

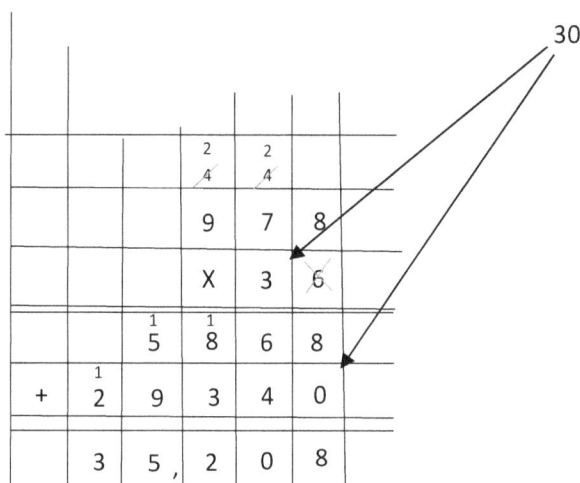

$$\begin{array}{ccccc}
 & & \overset{2}{\cancel{4}} & \overset{2}{\cancel{4}} & \\
 & & 9 & 7 & 8 \\
 & & X & 3 & \cancel{6} \\
\hline
 & \overset{1}{5} & \overset{1}{8} & 6 & 8 \\
+ & \overset{1}{2} & 9 & 3 & 4 & 0 \\
\hline
3 & 5, & 2 & 0 & 8
\end{array}$$

2) 129 X 94 =

$$\begin{array}{ccccc}
 & \overset{2}{\cancel{1}} & \overset{8}{\cancel{3}} & \\
 & 1 & 2 & 9 \\
 & X & 9 & \cancel{4} \\
\hline
 & 5 & 1 & 6 \\
+ & 1 & \overset{1}{1} & 6 & 1 & 0 \\
\hline
1 & 2, & 1 & 2 & 6
\end{array}$$

3) 456 X 78 =

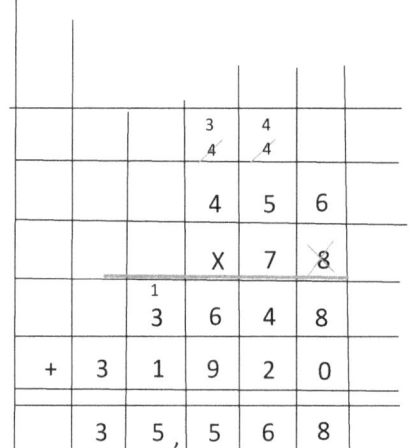

$$\begin{array}{ccccc}
 & \overset{3}{\cancel{4}} & \overset{4}{\cancel{4}} & \\
 & 4 & 5 & 6 \\
 & X & 7 & \cancel{8} \\
\hline
 & \overset{1}{3} & 6 & 4 & 8 \\
+ & 3 & 1 & 9 & 2 & 0 \\
\hline
3 & 5, & 5 & 6 & 8
\end{array}$$

Multiplication

C 3

Name _____

(1)

(2)   832 X 56 =

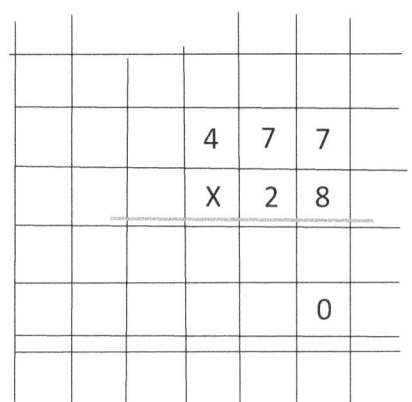

	4	7	7
	X	2	8
			0

(3)   563 X 49 =

(4)   1,009 X 75 =

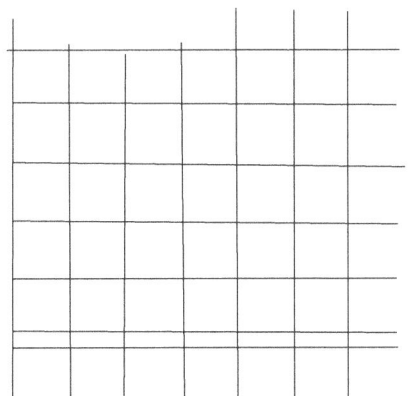

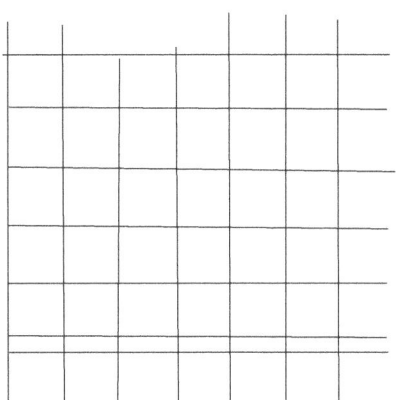

Multiplication

Name _____ Key _____

1

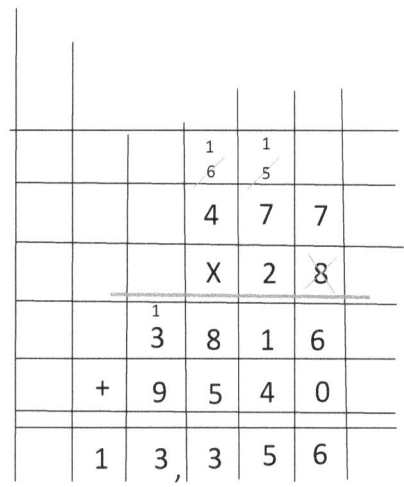

2   832 X 56 =

			1 6	1 5	
		4	7	7
		X	2	8
	1 3	8	1	6
+	9	5	4	0
1	3,	3	5	6

			1 1	1 1	
		8	3	2
		X	5	6
	1 4	9	9	2
+	4	1	6	0	0
	4	6,	5	9	2

3   563 X 49 =

4   1,009 X 75 =

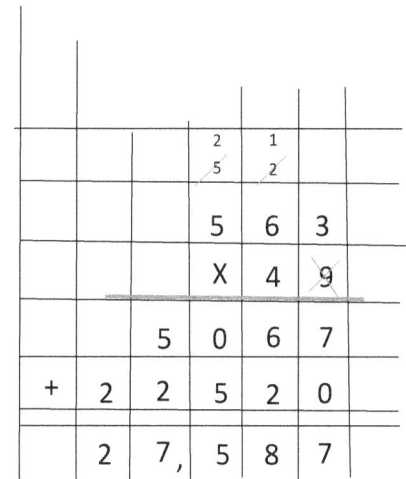

			2 5	1 2	
		5	6	3
		X	4	9
		5	0	6	7
+	2	2	5	2	0
	2	7,	5	8	7

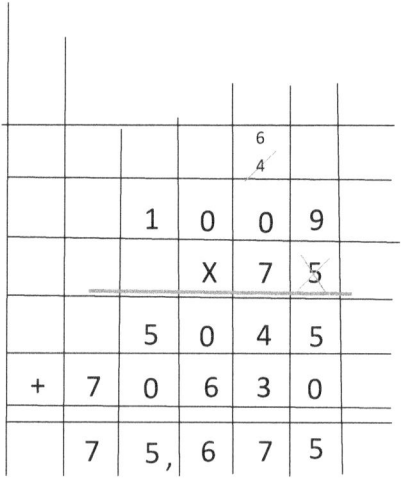

				6 4	
		1	0	0	9
			X	7	5
		5	0	4	5
+	7	0	6	3	0
	7	5,	6	7	5

Multiplication

C 4

Name _____

(1) 4026 X 36 =

(2) 824 X 79 =

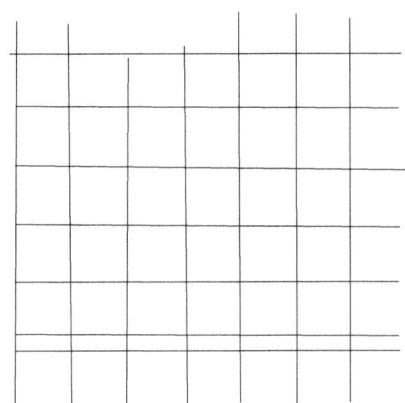

(3)

1,309 X 45 =

(4)

10,309 X 90 =

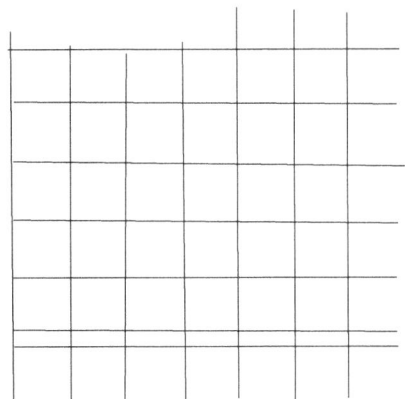

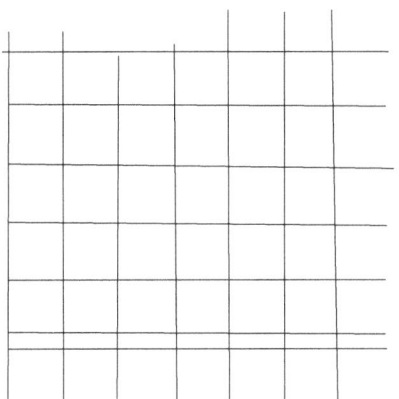

Multiplication

Name **Key**

① 4026 X 36 =

② 824 X 79 =

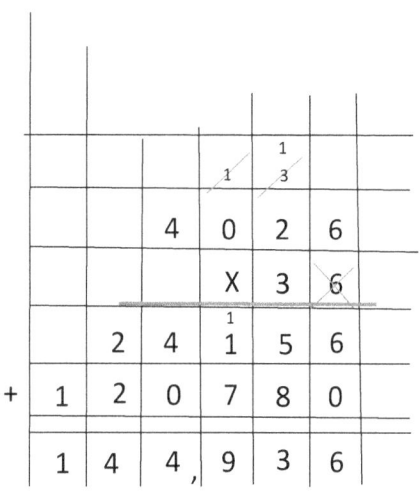

③ 1,309 X 45 =

④ 10,309 X 90 =

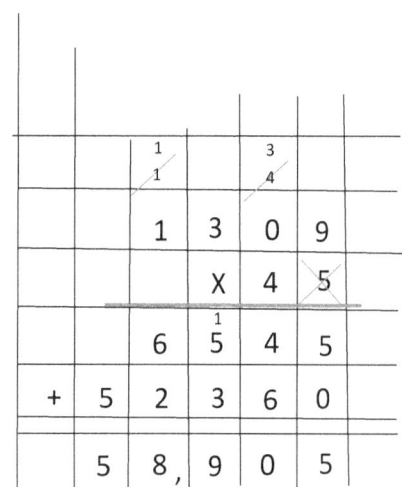

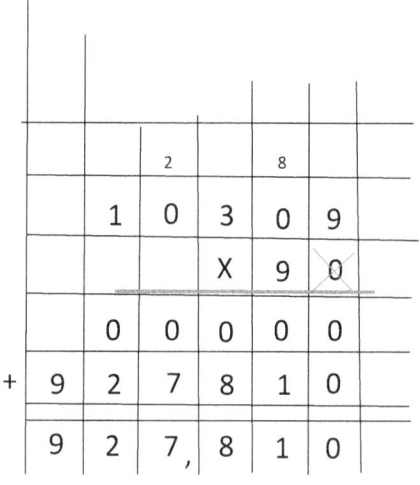

Multiplication

Name _____

1   326 X 28 =

2   547 X 82 =

3   931 X 76 =

4   235 X 45 =

5   702 X 16 =

6   923 X 70 =

Multiplication

Name __Key__

1

326 X 28 =

```
        2  1
          4
      3  2  6
   X     2  8
   ──────────
      1
      2  6  0  8
 +  6  5  2  0
 ──────────────
      9, 1  2  8
```

2

547 X 82 =

```
         3  5
            1
      5  4  7
   X     8  2
   ──────────
         1
   1  0  9  4  0
+  4  3  7  6  0
────────────────
   4  4, 8  5  4
```

3

931 X 76 =

```
         2
         1
      9  3  1
   X     7  6
   ──────────
      5  5  8  6  0
         1
+  6  5  1  7  0
────────────────
   7  0, 7  5  6
```

4

235 X 45 =

```
    1  2
    1  2
    2 3 5
  X 4 5
  ────────
    1 1 7 5
 + 9 4 0 0
 ──────────
  1 0, 5 7 5
```

5

702 X 16 =

```
       1
    7 0 2
  X 1 6
  ────────
    4 2 1 2
 + 7 0 2 0
 ──────────
  1 1, 2 3 2
```

6

923 X 70 =

```
   1  2
    9 2 3
  X 7 0
  ────────
    0 0 0
 + 6 4 6 1 0
 ──────────
  6 4, 6 1 0
```

Division With Factors

Name _____

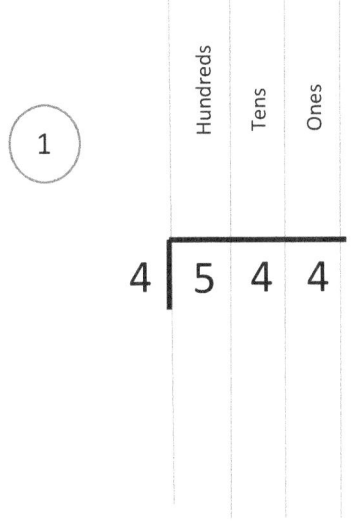

1

	Hundreds	Tens	Ones
4)	5	4	4

4
8
12
16
20
24
28
32
36
40

2

6)	3	4	2

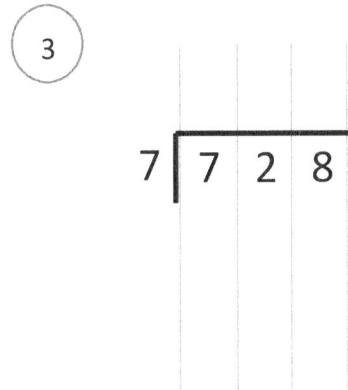

3

7)	7	2	8

4

8)	7	8	4

Division With Factors

D 1

Name Key

①

	Hundreds	Tens	Ones
	1	3	6
4	5	4	4
	-4	↓	
	1	4	
	-1	2	↓
		2	4
		-2	4
			0

(4), 8, (12), 16, 20, (24), 28, 32, 36, 40

②

	0	5	7
6	3	4	2
	-3	0	↓
		4	2
		-4	2
			0

6, 12, 18, 24, (30), 36, (42), 48, 54, 60

③

	1	0	4
7	7	2	8
	-7	↓	↓
	0	2	8
		-2	8
			0

(7), 14, 21, (28), 35, 42, 49, 56, 63, 70

④

	0	9	8
8	7	8	4
	-7	2	↓
		6	4
		-6	4
			0

8, 16, 24, 32, 40, 48, 56, (64), (72), 80

Division With Factors

Name _____

1

Hundreds	Tens	Ones

13 | 5 5 9

13
26
39
52
65
78
91
104
117
130

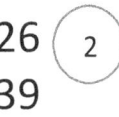

2

3 | 6 2 7

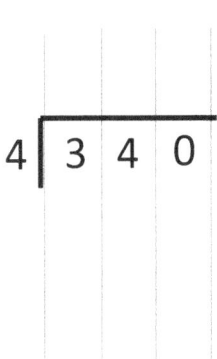

3

4 | 3 4 0

4

9 | 7 2 9

Division With Factors

Name **Key**

1

	Hundreds	Tens	Ones
	0	4	3
13	5	5	9
	-5	2	↓
		3	9
	-	3	9
			0

13
26
(39)
(52)
65
78
91
104
117
130

2

	Hundreds	Tens	Ones
	2	0	9
3	6	2	7
	-6	↓	↓
	0	2	7
	-	2	7
			0

3
(6)
9
12
15
18
21
24
(27)
30

3

	Hundreds	Tens	Ones
	0	8	5
4	3	4	0
	-3	2	↓
		2	0
	-	2	0
			0

4
8
12
16
(20)
24
28
(32)
36
40

4

	Hundreds	Tens	Ones
	0	8	1
9	7	2	9
	-7	2	↓
		0	9
	-		9
			0

(09)
18
27
36
45
54
63
(72)
81
90

Division With Factors

Name _____

| | Hundreds | Tens | Ones |

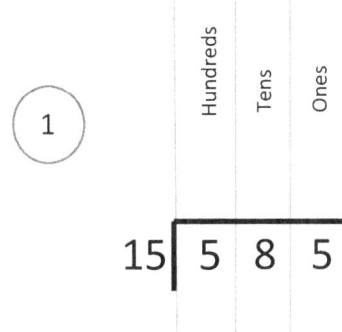

1

15 | 5 | 8 | 5

15
30
45
60
75
90
105
120
135
150

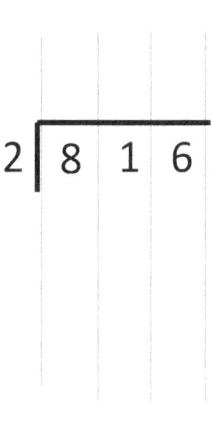

2

2 | 8 | 1 | 6

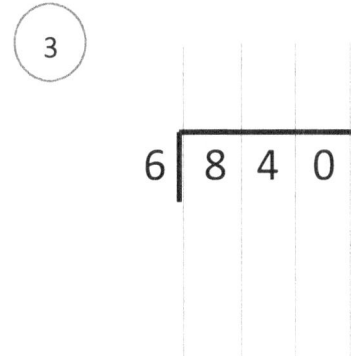

3

6 | 8 | 4 | 0

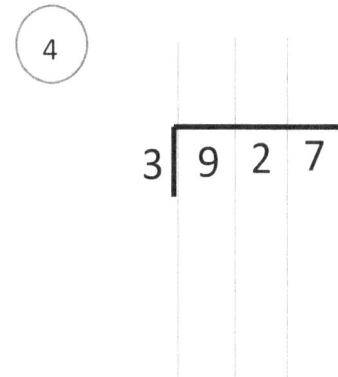

4

3 | 9 | 2 | 7

Yellow Pencil Mathematics ™

Division With Factors

Name ___Key___

1

	Hundreds	Tens	Ones
	0	3	9
15	5	8	5
	-4	5	↓
	1	3	5
	- 1	3	5
			0

15
30
(45)
60
75
90
105
120
(135)
150

2

	Hundreds	Tens	Ones
	4	0	8
2	8	1	6
	-8	↓	↓
	0	1	6
	-	1	6
			0

2
4
6
(8)
10
12
14
(16)
18
20

3

	Hundreds	Tens	Ones
	1	4	0
6	8	4	0
	-6	↓	↓
	2	4	
	- 2	4	↓
		0	0

(6)
12
18
(24)
30
36
42
48
54
60

4

	Hundreds	Tens	Ones
	3	0	9
3	9	2	7
	-9	↓	↓
	0	2	7
	-	2	7
			0

3
6
(9)
12
15
18
21
24
(27)
30

Yellow Pencil Mathematics ™

①   6 | 4 8 2 4 .

②   7 | 7 0 6 3

③   8 | 1 8 7 2

④   9 | 1 6 8 3

Division With Factors

Name _____Key_____

1

	Thousands	Hundreds	Tens	Ones
		0	8	0	4
6		4	8	2	4
		-4	8		
		0	2	4
			-2	4
				0

6
12
18
(24)
30
36
42
48
54
60

2

	Thousands	Hundreds	Tens	Ones
		1	0	0	9
7		7	0	6	3
		-7			
		0	0	6	3
				-6	3
				0

(7)
14
21
28
35
42
49
56
(63)
70

3

	Thousands	Hundreds	Tens	Ones
		0	2	3	4
8		1	8	7	2
		-1	6		
			2	7	
			-2	4	
				3	2
			-3	3	2
				0

8
(16)
(24)
(32)
40
48
56
64
72
80

4

	Thousands	Hundreds	Tens	Ones
		0	1	8	7
9		1	6	8	3
		-9			
			7	8	
			-7	2	
				6	3
			-6	3
				0

(09)
18
27
38
45
54
(63)
(72)
81
90

Yellow Pencil Mathematics ™

Division With Factors

①

Thousands Hundreds Tens Ones

13 | 1 1 5 7 .

②

15 | 1 5 3 0

③

17 | 2 4 1 4

④

19 | 1 4 4 4

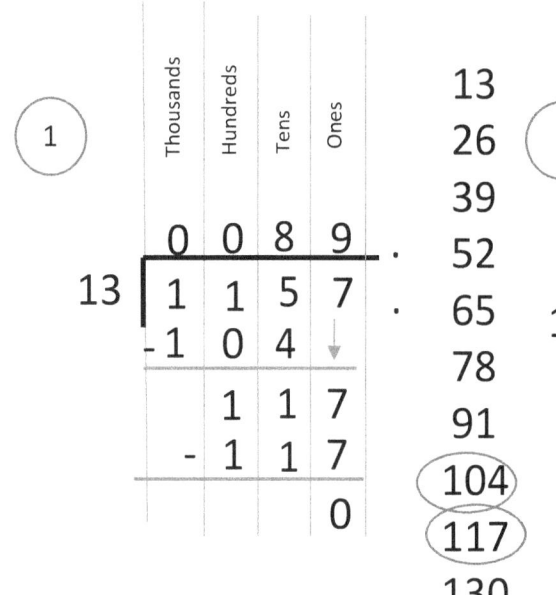

1

Thousands	Hundreds	Tens	Ones
0	0	8	9 .

13)

| 1 | 1 | 5 | 7 . |
| -1 | 0 | 4 | ↓ |

| | 1 | 1 | 7 |
| - | 1 | 1 | 7 |

| | | | 0 |

13
26
39
52
65
78
91
(104)
(117)
130

2

Thousands	Hundreds	Tens	Ones
0	1	0	2

15)

| 1 | 5 | 3 | 0 |
| -1 | 5 | ↓ | ↓ |

| | 0 | 3 | 0 |
| | - | 3 | 0 |

| | | | 0 |

(15)
(30)
45
60
75
90
105
120
135
150

3

Thousands	Hundreds	Tens	Ones
0	1	4	2

17)

| ₁2̶ | ₁4 | 1 | 4 |
| -1 | 7 | ↓ | ↓ |

| | 7 | 1 | |
| | - 6 | 8 | ↓ |

| | | 3 | 4 |
| | | - 3 | 4 |

| | | | 0 |

(17)
(34)
51
(68)
85
102
119
136
153
170

4

Thousands	Hundreds	Tens	Ones
0	0	7	6

19)

| 1 | 4 | 4 | 4 |
| -1 | 3 | 3 | ↓ |

| | 1 | 1 | 4 |
| | - 1 | 1 | 4 |

| | | | 0 |

19
38
57
76
95
(114)
(133)
152
171
190

56

Yellow Pencil Mathematics ™

Place Value

E 1   Name _____

Hundred Thousands	Ten Thousands	Thousands	Hundreds	Tens	Ones	And	Tenths	Hundredths	Thousandths
						.			

1.   In the number 123,456.789 the number eight is in what place value? _____

2.   In the number 123,456.789 the number two is in what place value? _____

3.   In the number 765.34 the number five is in what place value? _____

4.   In the number 123,456 the number one is in what place value? _____

5.   In the number 56.712 the number one is in what place value? _____

6.   In the number 123,456.789 the number three is in what place value? _____

7.   In the number 123,456.789 the number eight is in what place value? _____

8.   In the number 123,456.789 the number one is in what place value? _____

9.   In the number 1,987.23 the number eight is in what place value? _____

10.   In the number 345 where does the decimal go? _____

11.   In the number 1,987 the number eight is in what place value? _____

12.   In the number 1,987.23 the number one is in what place value? _____

13.   In the number 1,987.236 the number six is in what place value? _____

14.   In the number 1,439,345 where does the decimal go? _____

15.   In the number 0.2309 the number two is in what place value? _____

Place Value

Name _____**Key**_____

Hundred Thousands	Ten Thousands	Thousands	Hundreds	Tens	Ones	And	Tenths	Hundredths	Thousandths
						.			

1. In the number 123,456.789 the number eight is in what place value? Hundredths

2. In the number 123,456.789 the number two is in what place value? Ten Thousands

3. In the number 765.34 the number five is in what place value? Ones

4. In the number 123,456 the number one is in what place value? Hundred Thousands

5. In the number 56.712 the number one is in what place value? Hundredths

6. In the number 123,456.789 the number three is in what place value? Thousands

7. In the number 123,456.789 the number eight is in what place value? Hundredths

8. In the number 123,456.789 the number one is in what place value? Hundred Thousands

9. In the number 1,987.23 the number eight is in what place value? Tens

10. In the number 345 where does the decimal go? Behind the five Behind the ones place

11. In the number 1,987 the number eight is in what place value? Tens

12. In the number 1,987.23 the number one is in what place value? Thousands

13. In the number 1,987.236 the number six is in what place value? Thousandths

14. In the number 1,439,345 where does the decimal go? Behind the five Behind the ones place

15. In the number 0.2309 the number two is in what place value? Tenths

Place Value

Name _____

					And				
					.				

Tenths   Tens

Hundred Thousands

Hundredths

Ones   Hundreds

Ten Thousands

Thousandths

Thousands

1. In the number 123,456.789 the number three is in what place value? _____

2. In the number 123,456.789 the number seven is in what place value? _____

3. In the number 765.34 the number six is in what place value? _____

4. In the number 123.456 the number five is in what place value? _____

5. In the number 56.712 the number seven is in what place value? _____

6. In the number 123,456.789 the number six is in what place value? _____

7. In the number 123,456.789 the number nine is in what place value? _____

8. In the number 123,456.789 the number one is in what place value? _____

9. In the number 1,987.23 the number two is in what place value? _____

10. In the number 945 where does the decimal go? _____

11. In the number 1,987.23 the number nine is in what place value? _____

12. In the number 1,987.23 the number three is in what place value? _____

13. In the number 1,987.236 the number six is in what place value? _____

14. In the number 00987 where does the decimal go? _____

15. In the number 0.2309 the number three is in what place value? _____

Place Value

Name _____ **Key**

Hundred Thousands	Ten Thousands	Thousands	Hundreds	Tens	Ones	And	Tenths	Hundredths	Thousandths
						.			

1. In the number 123,456.789 the number three is in what place value? ___Thousands___

2. In the number 123,456.789 the number seven is in what place value? ___Tenths___

3. In the number 765.34 the number six is in what place value? ___Tens___

4. In the number 123.456 the number five is in what place value? ___Hundredths___

5. In the number 56.712 the number seven is in what place value? ___Tenths___

6. In the number 123,456.789 the number six is in what place value? ___Ones___

7. In the number 123,456.789 the number nine is in what place value? ___Thousandths___

8. In the number 123,456.789 the number one is in what place value? Hundred Thousands

9. In the number 1,987.23 the number two is in what place value? ___Tenths___

10. In the number 945 where does the decimal go? Behind the five or Behind the ones place

11. In the number 1,987.23 the number nine is in what place value? ___Hundreds___

12. In the number 1,987.23 the number three is in what place value? ___Hundredths___

13. In the number 1,987.236 the number six is in what place value? ___Thousandths___

14. In the number 00987 where does the decimal go? Behind the seven or Behind the ones place

15. In the number 0.2309 the number three is in what place value? ___Hundredths___

Writing Numbers in Expanded Form

F 1

Name _____

Example of Expanded Form 3,250.5 = 3,000 + 200 + 50 + 0.5

Place the following numbers in the charts; then list them in Expanded Form

30,543

(1)

Ten Thousands	Thousands	Hundreds	Tens	Ones	And	Tenths	Hundredths
					.		

954.38

(2)

Ten Thousands	Thousands	Hundreds	Tens	Ones	And	Tenths	Hundredths
					.		

63,001.3

(3)

Ten Thousands	Thousands	Hundreds	Tens	Ones	And	Tenths	Hundredths
					.		

2,035.2

(4)

Ten Thousands	Thousands	Hundreds	Tens	Ones	And	Tenths	Hundredths
					.		

Writing Numbers in Expanded Form

Name _____ **Key**

Example of Expanded Form 3,250.5 = 3,000 + 200 + 50 + 0.5

Place the following numbers in the charts; then list them in Expanded Form

30,543

Ten Thousands	Thousands	Hundreds	Tens	Ones	And	Tenths	Hundredths
3	0	0	0	0	.		
		5	0	0	.		
			4	0	.		
				3	.		

30,000 + 500 + 40 + 3

954.38

Ten Thousands	Thousands	Hundreds	Tens	Ones	And	Tenths	Hundredths
		9	0	0	.		
			5	0	.		
				4	.		
				0	.	3	
				0	.	0	8

900 + 50 + 4 + 0.3 + 0.08

63,001.3

Ten Thousands	Thousands	Hundreds	Tens	Ones	And	Tenths	Hundredths
6	0	0	0	0	.		
	3	0	0	0	.		
				1	.		
				0	.	3	

60,000 + 3,000 + 1 + 0.3

2,035.2

Ten Thousands	Thousands	Hundreds	Tens	Ones	And	Tenths	Hundredths
	2	0	0	0	.		
			3	0	.		
				5	.		
				0	.	2	

2000 + 30 + 5 + 0.2

Writing Numbers in Expanded Form

Name _____

Place the following numbers in the charts; then list them in Expanded Form

29,000.69

(1)

Ten Thousands	Thousands	Hundreds	Tens	Ones	And	Tenths	Hundredths
					.		

9,033.8

(2)

Ten Thousands	Thousands	Hundreds	Tens	Ones	And	Tenths	Hundredths
					.		

0.96

(3)

Ten Thousands	Thousands	Hundreds	Tens	Ones	And	Tenths	Hundredths
					.		

568.24

(4)

Ten Thousands	Thousands	Hundreds	Tens	Ones	And	Tenths	Hundredths
					.		

Writing Numbers in Expanded Form

Name _____ **Key** _____

Place the following numbers in the charts; then list them in Expanded Form

29,000.69

(1)

Ten Thousands	Thousands	Hundreds	Tens	Ones	And	Tenths	Hundredths
2	0	0	0	0	.		
	9	0	0	0	.		
				0	.	6	
				0	.	0	9

20,000 + 9,000 + 0.6 + 0.09

9,033.8

(2)

Ten Thousands	Thousands	Hundreds	Tens	Ones	And	Tenths	Hundredths
	9	0	0	0	.		
			3	0	.		
				3	.		
				0	.	8	

9,000 + 30 + 3 + 0.8

0.96

(3)

Ten Thousands	Thousands	Hundreds	Tens	Ones	And	Tenths	Hundredths
				0	.	9	
				0	.	0	6

0.9 + 0.06

568.24

(4)

Ten Thousands	Thousands	Hundreds	Tens	Ones	And	Tenths	Hundredths
		5	0	0	.		
			6	0	.		
				8	.		
				0	.	2	
				0	.	0	4

500 + 60 + 8 + 0.2 + 0.04

Writing Numbers in Expanded Notation

F 3

Name _____

Example of Expanded Form 3,250.5 = (3 X 1000) + (2 X 100) + (5 X 10) + (5 X 0.1)

Place the following numbers in the charts; then list them in Expanded Notation

30,543

(1)

Ten Thousands	Thousands	Hundreds	Tens	Ones	And	Tenths	Hundredths
					.		

(___ X ___) + (___ X ___) + (___ X ___) + (___ X ___)

954.3

(2)

Ten Thousands	Thousands	Hundreds	Tens	Ones	And	Tenths	Hundredths
					.		

63,001.3

(3)

Ten Thousands	Thousands	Hundreds	Tens	Ones	And	Tenths	Hundredths
					.		

2,035.27

(4)

Ten Thousands	Thousands	Hundreds	Tens	Ones	And	Tenths	Hundredths
					.		

Writing Numbers in Expanded Notation

F 3

Name ____Key____

Example of Expanded Notation 3,250.5 = (3 X 1000) + (2 X 100) + (5 X 10) + (5 X 0.1)

Place the following numbers in the charts; then list them in Expanded Notation

30,543

(1)

Ten Thousands	Thousands	Hundreds	Tens	Ones	And	Tenths	Hundredths
3	0	0	0	0	.		
		5	0	0	.		
			4	0	.		
				3	.		

(3 X 10,000) + (5 X 100) + (4 X 10) + (3 X 1)

954.3

(2)

Ten Thousands	Thousands	Hundreds	Tens	Ones	And	Tenths	Hundredths
		9	0	0	.		
			5	0	.		
				4	.		
				0	.	3	

(9 X 100) + (5 X 10) + (4 X 1) + (3 X 0.1)

63,001.3

(3)

Ten Thousands	Thousands	Hundreds	Tens	Ones	And	Tenths	Hundredths
6	0	0	0	0	.		
	3	0	0	0	.		
				1	.		
				0	.	3	

(6 X 10,000) + (3 X 1,000) + (1 X 1) + (3 X 0.1)

2,035.07

(4)

Ten Thousands	Thousands	Hundreds	Tens	Ones	And	Tenths	Hundredths
	2	0	0	0	.		
			3	0	.		
				5	.		
				0	.	0	7

(2 X 1,000) + (3 X 10) + (5 X 1) + (7 X 0.01)

Writing Numbers in Expanded Notation

F 4

Name _____

Example of Expanded Form 3,250.5 = (3 X 1000) + (2 X 100) + (5 X 10) + (5 X 0.1)

Place the following numbers in the charts; then list them in Expanded Notation

29.69

(1)

Ten Thousands	Thousands	Hundreds	Tens	Ones	And	Tenths	Hundredths
					.		

(X   _____) + (X   _____) + (X   _____) + (X   _____)

9,022.8

(2)

Ten Thousands	Thousands	Hundreds	Tens	Ones	And	Tenths	Hundredths
					.		

0.96

(3)

Ten Thousands	Thousands	Hundreds	Tens	Ones	And	Tenths	Hundredths
					.		

560.24

(4)

Ten Thousands	Thousands	Hundreds	Tens	Ones	And	Tenths	Hundredths
					.		

Writing Numbers in Expanded Notation

F 4

Name __Key__

Example of Expanded Notation 3,250.5 = (3 X 1000) + (2 X 100) + (5 X 10) + (5 X 0.1)

Place the following numbers in the charts; then list them in Expanded Notation

29.69

(1)

Ten Thousands	Thousands	Hundreds	Tens	Ones	And	Tenths	Hundredths
			2	0	.		
				9	.		
				0	.	6	
				0	.	0	9

(2 X 10) + (9 X 1) + (6 X 0.1) + (9 X 0.01)

9,022.8

(2)

Ten Thousands	Thousands	Hundreds	Tens	Ones	And	Tenths	Hundredths
	9	0	0	0	.		
			2	0	.		
				2	.		
				0	.	8	

(9 X 1,000) + (2 X 10) + (2 X 1) + (8 X 0.1)

0.96

(3)

Ten Thousands	Thousands	Hundreds	Tens	Ones	And	Tenths	Hundredths
				0	.	9	
				0	.	0	6

(9 X 0.1) + (6 X 0.01)

560.24

(4)

Ten Thousands	Thousands	Hundreds	Tens	Ones	And	Tenths	Hundredths
		5	0	0	.		
			6	0	.		
				0	.	2	
				0	.	0	4

(5 X 100) + (6 X 10) + (2 X 0.1) + (4 X 0.01)

G 1   Number Lines

Place a point to mark the given number on the number line.

(70)

0   100

(35)

0   100

(28)

0   100

(53)

0   100

(3)

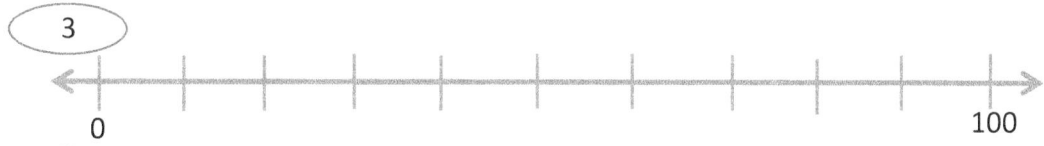

0   100

(300)

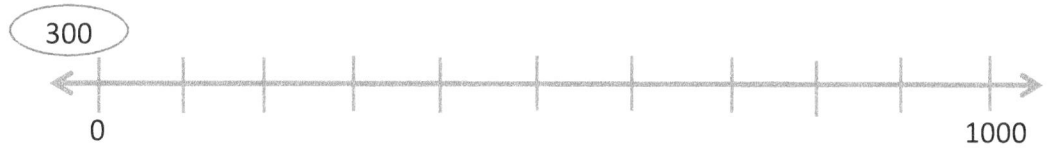

0   1000

(850)

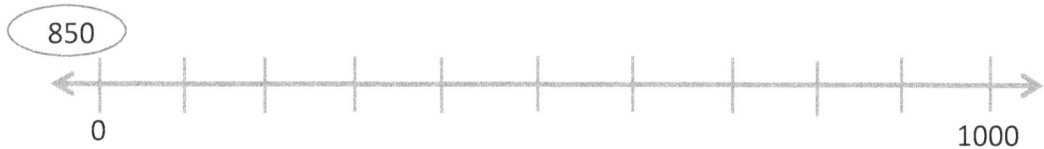

0   1000

(225)

0   1000

69

Yellow Pencil Mathematics™

Number Lines

Name _____ Key _____

Place a point to mark the given number on the number line.

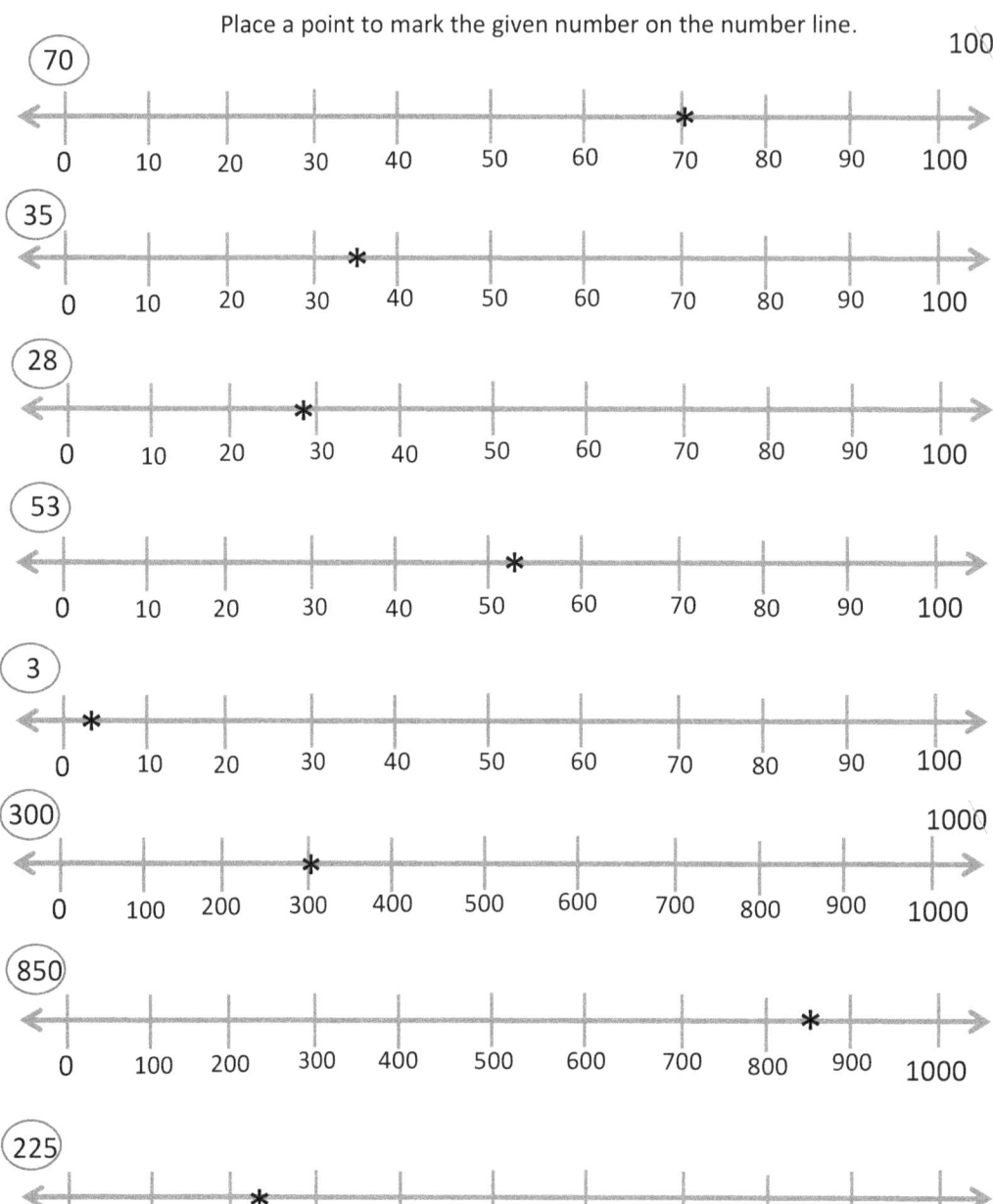

Number Lines

Name _____

Place the given number on the number line.

65
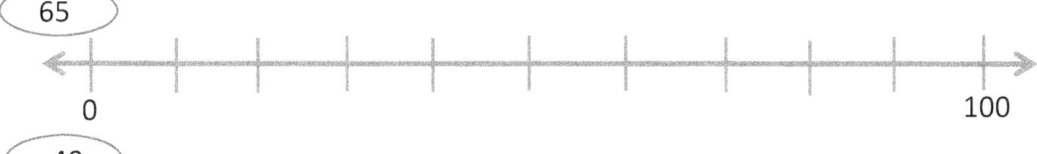
0   100

40

0   200

300
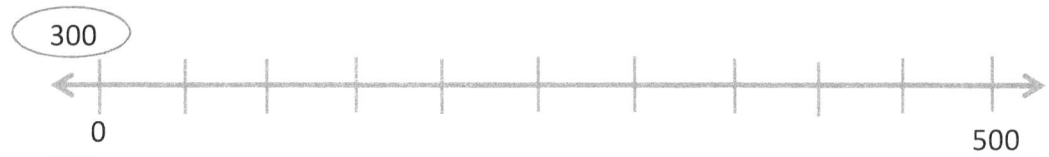
0   500

800
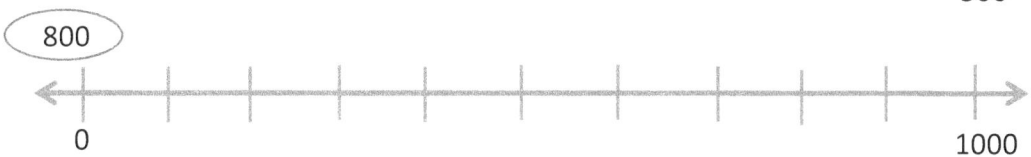
0   1000

15
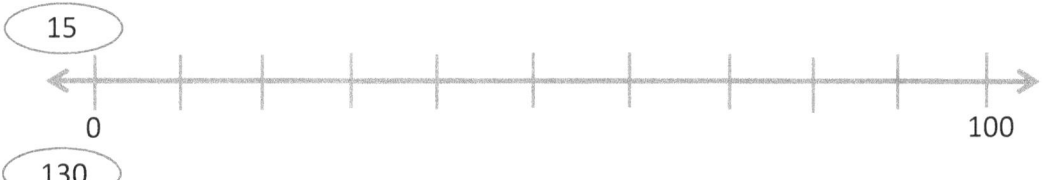
0   100

130

0   200

150

0   500

450

0   1000

Number Lines

Name _____ **Key** _____

Place the given number on the number line.

(65)   100

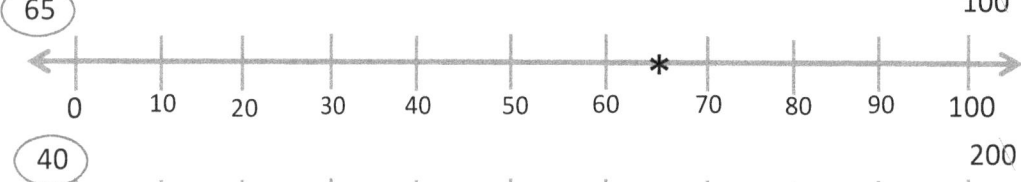

(40)   200

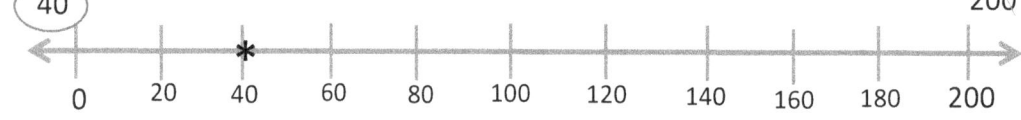

(300)   500

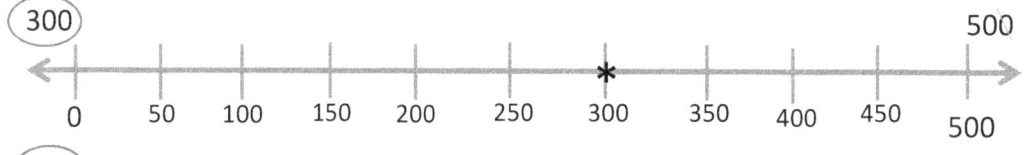

(800)   1000

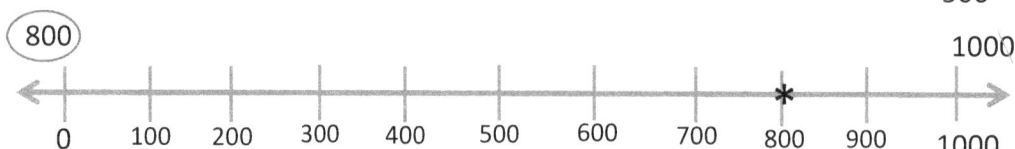

(15)

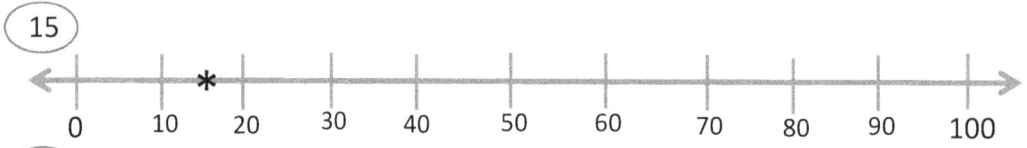

(130)

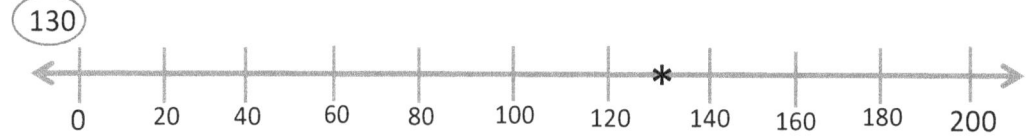

(150)

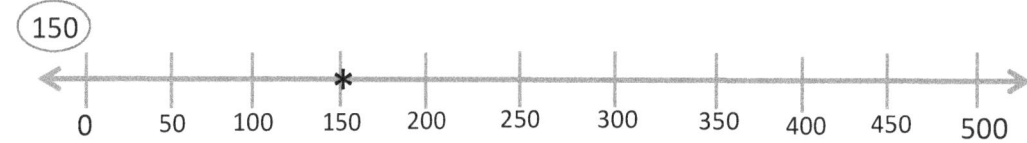

(450)

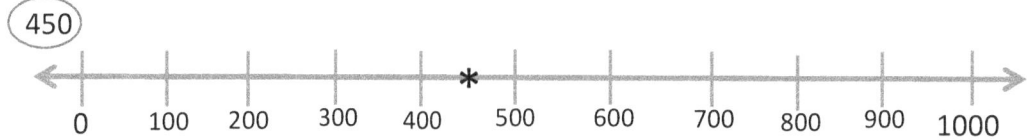

Number Lines

Name _____

Place the given number on the number line.

72

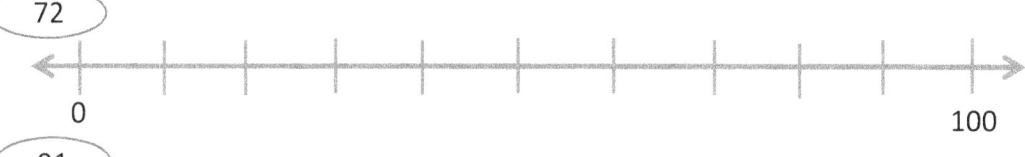

0   100

91

0   200

312

0   500

327

0   1000

32

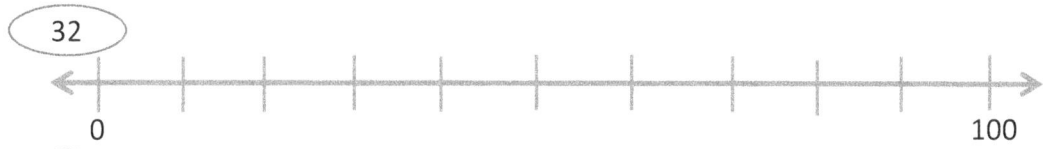

0   100

107

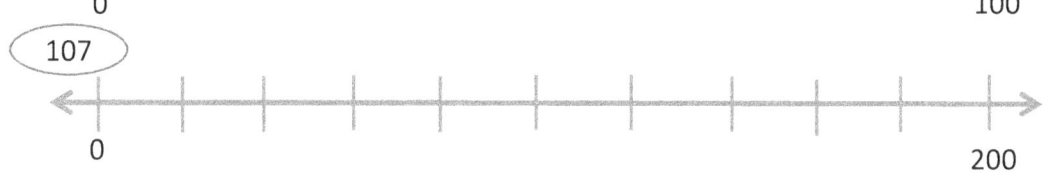

0   200

82

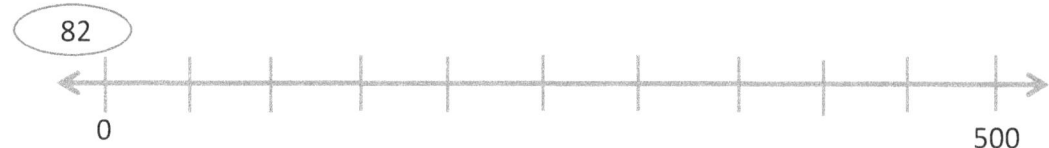

0   500

622

0   1000

Number Lines

Name _____ **Key** _____

Place the given number on the number line.

(72)

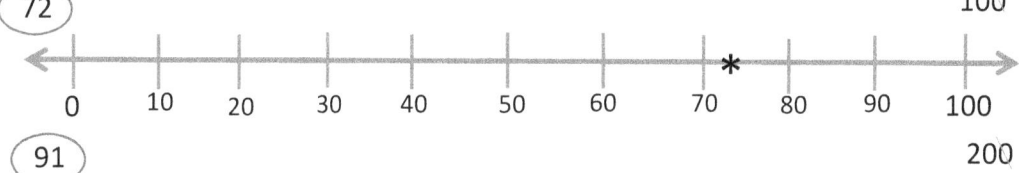

(91)

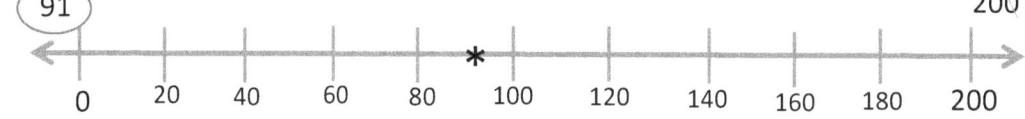

(312)

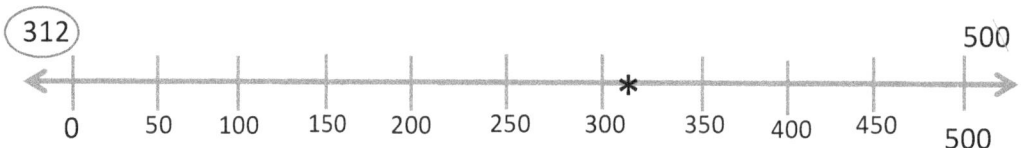

(327)

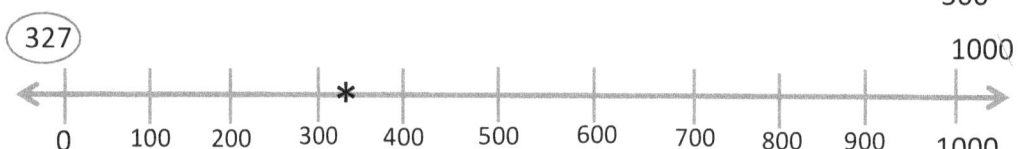

(32)

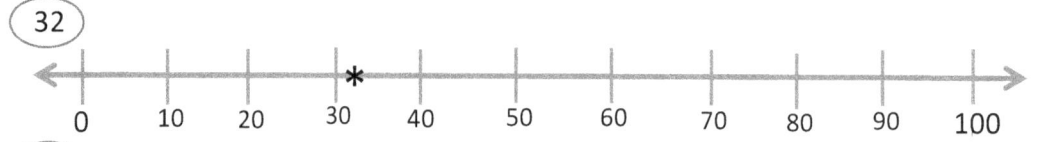

(107)

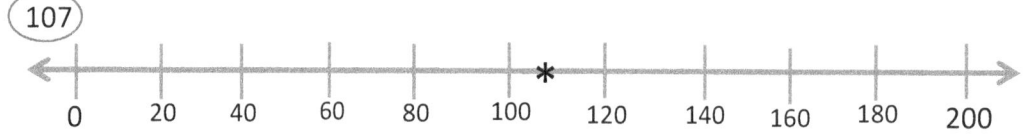

(82)

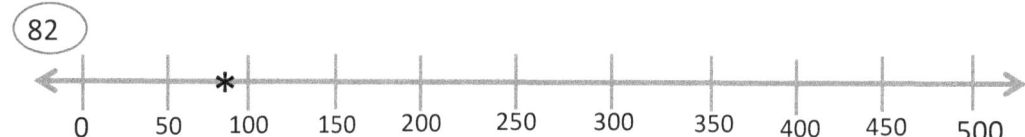

(622)

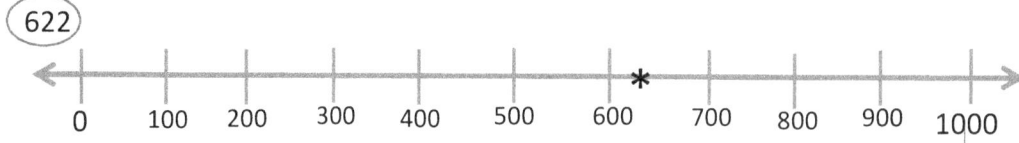

Number Lines

Name _____

Place a point for the given number on the number line.

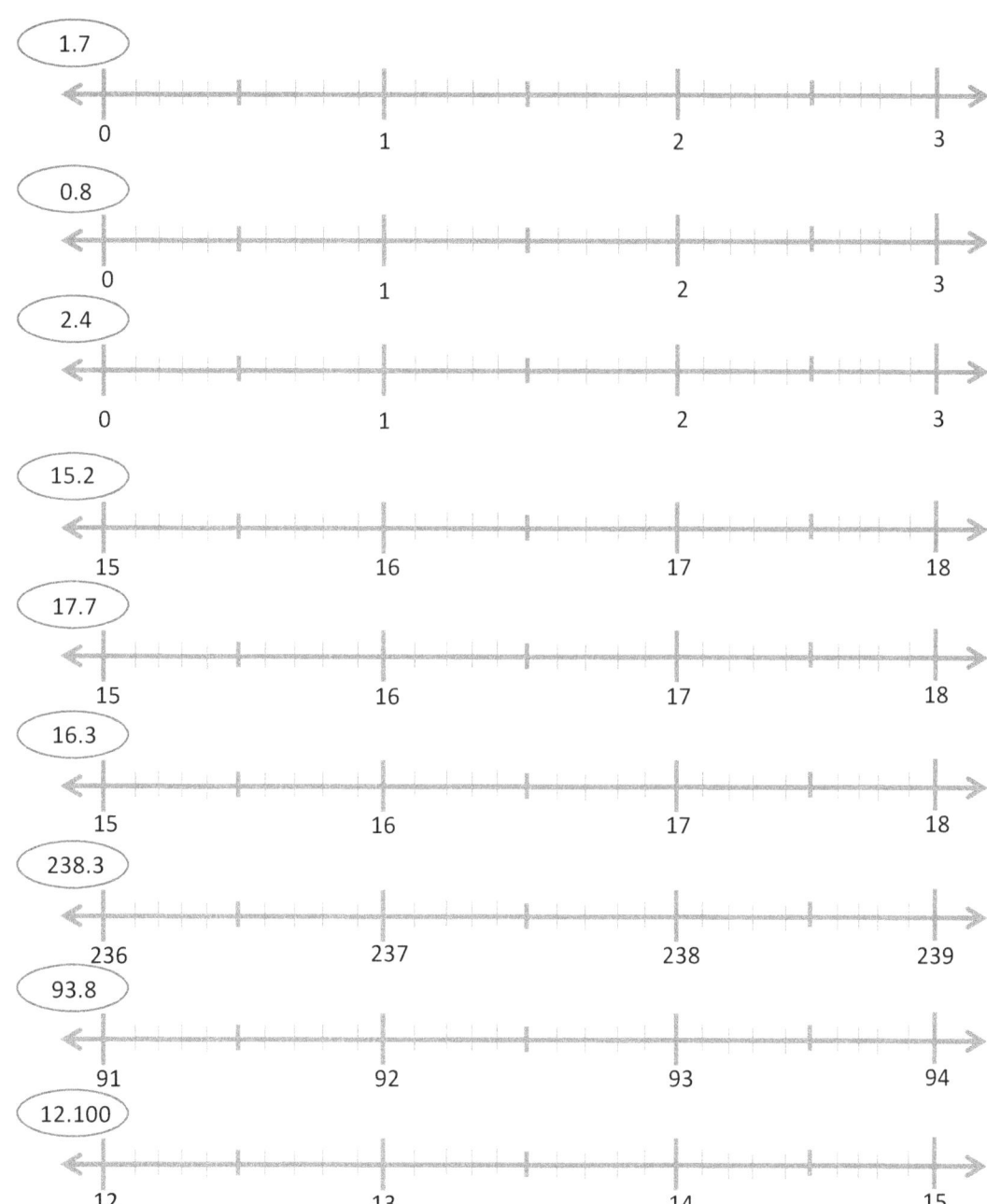

Number Lines

Name _____**Key**_____

Place a point for the given number on the number line.

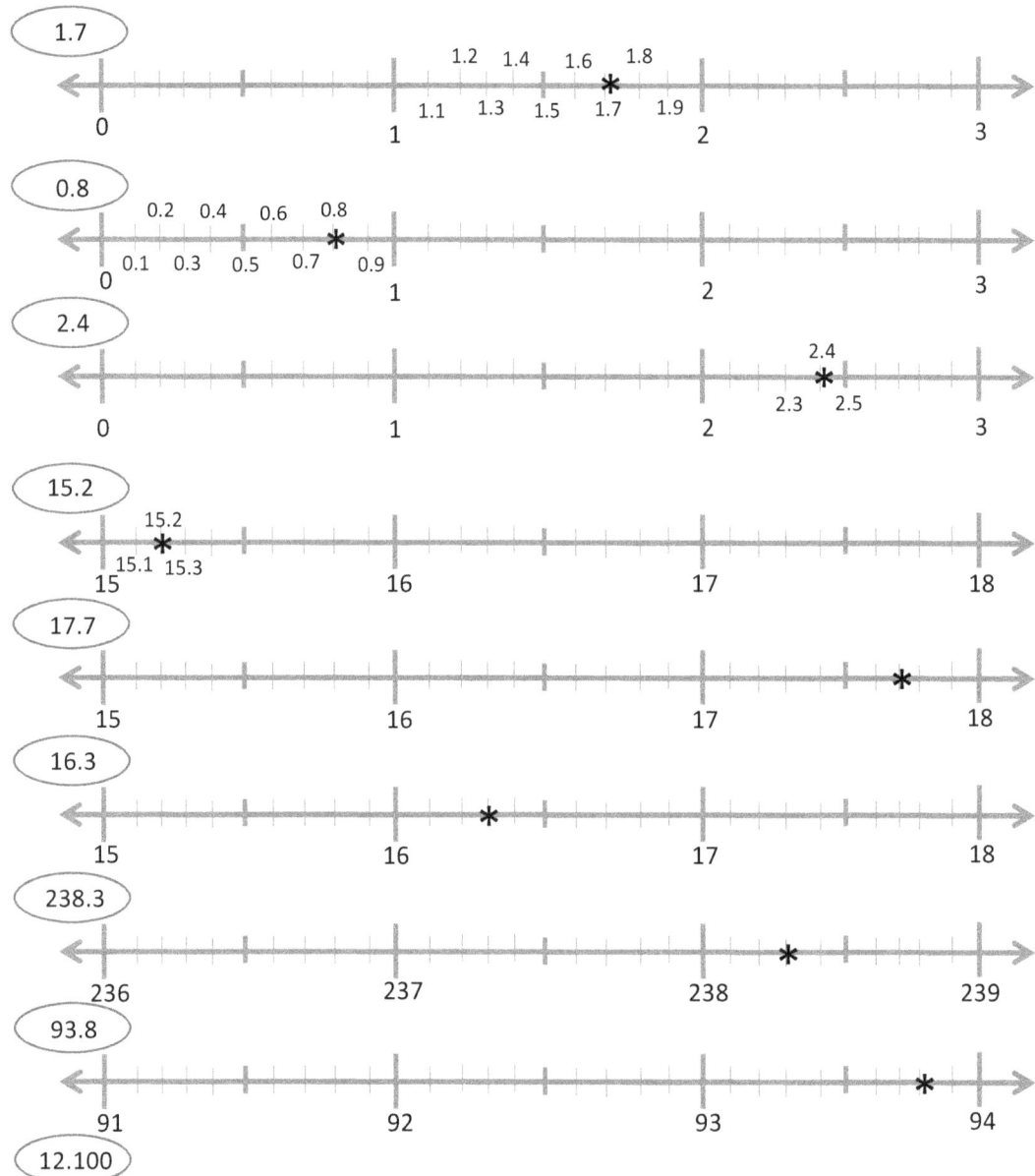

Number Lines

Name _____

Place a point for the given number on the number line.

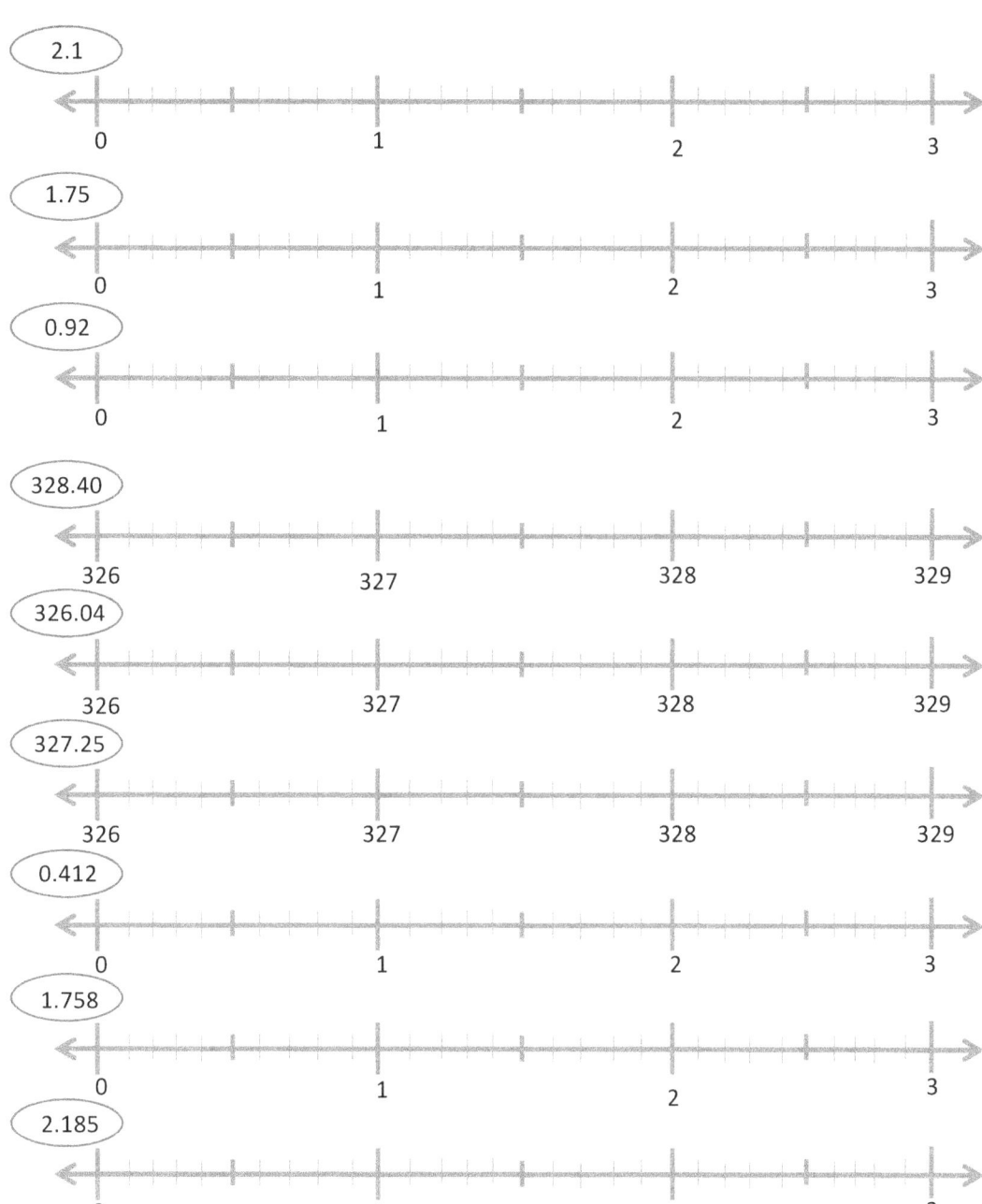

Number Lines

Name _____**Key**_____

Place a point for the given number on the number line.

2.1

1.75

0.92

328.40

326.04

327.25

0.412

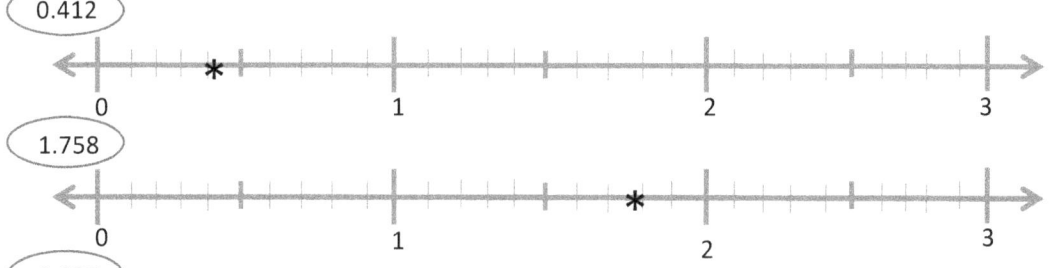

1.758

2.185

Number Lines

GH 1

Name _____

Place the given number on the number line. Then **estimate** to the nearest ones place.

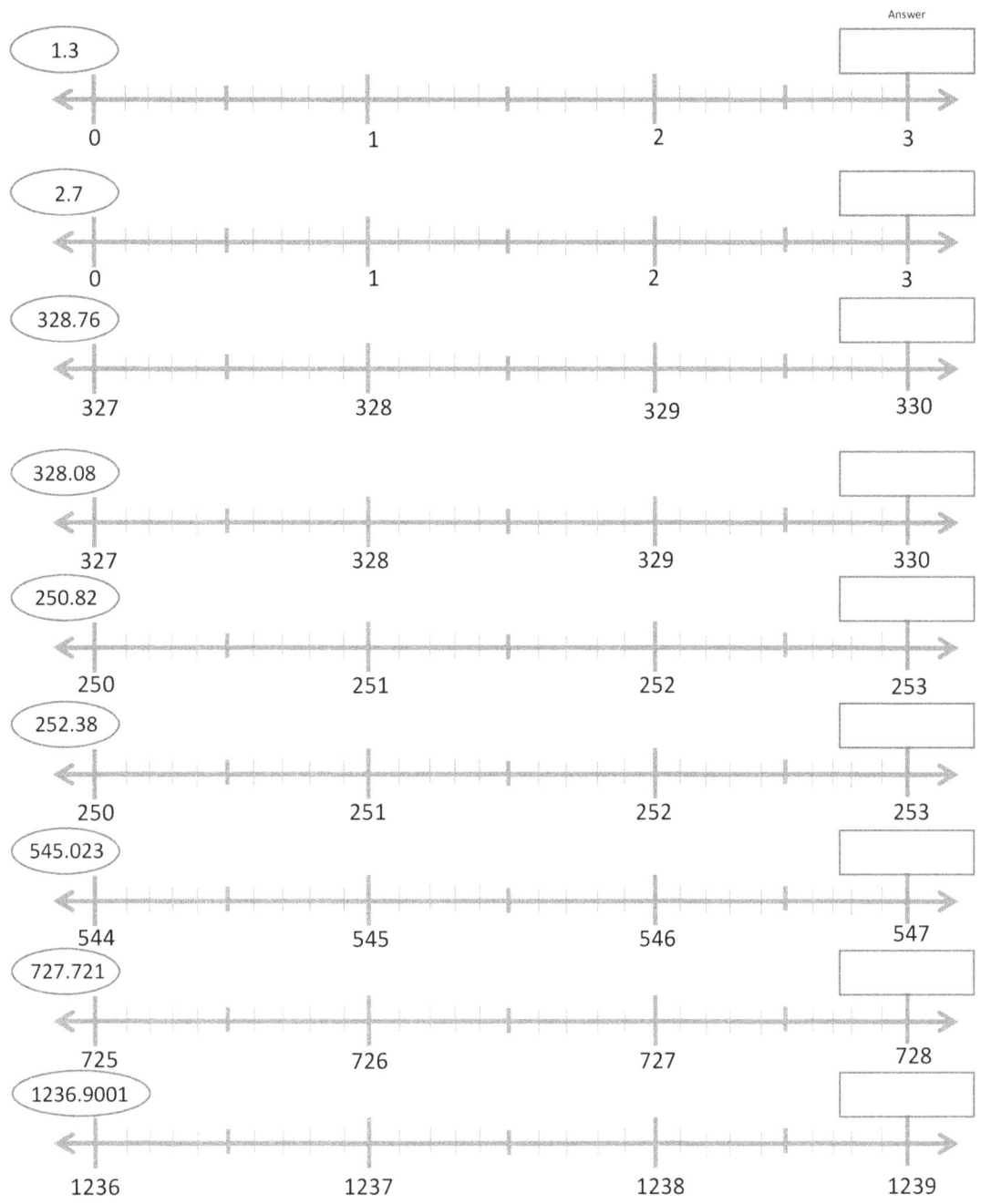

Yellow Pencil Mathematics™

GH 1

Name ___Key___

Place the given number on the number line. Then **estimate** to the nearest ones place.

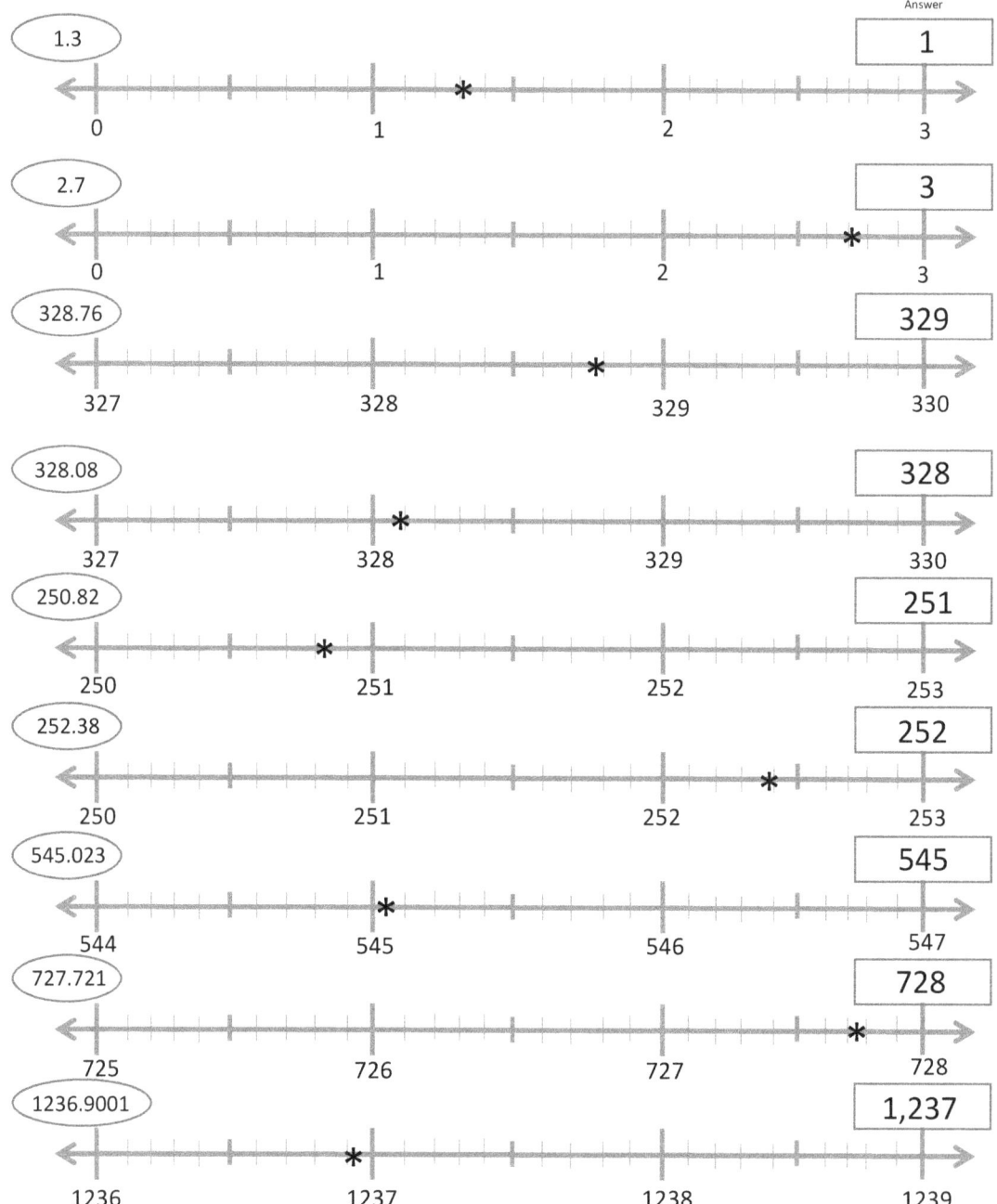

Number Lines

Name _____

Place the given number on the number line. Then **estimate**
to the nearest place value indicated.

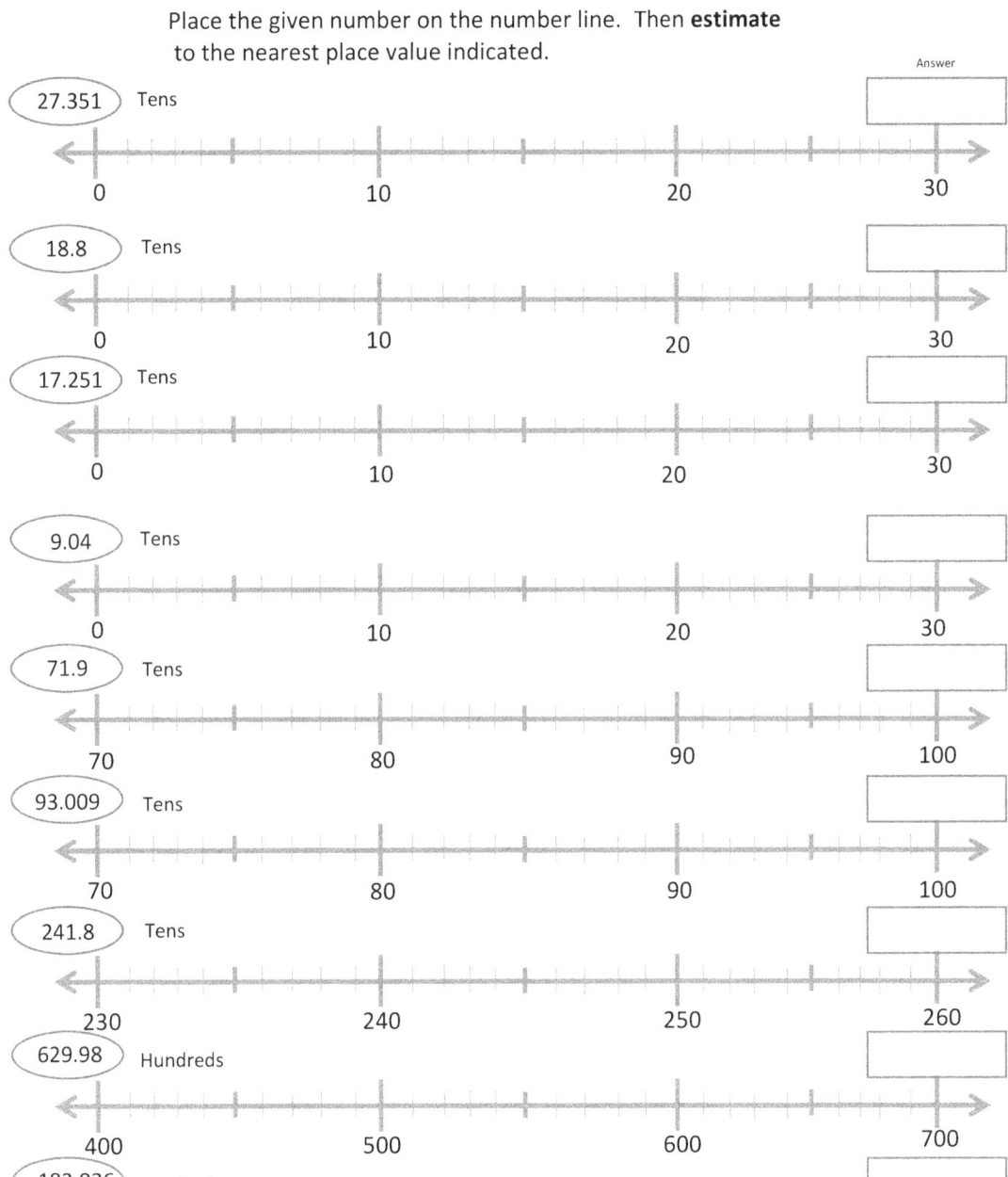

Number Lines

Name ___**Key**___

Place the given number on the number line. Then **estimate**
to the nearest place value indicated.

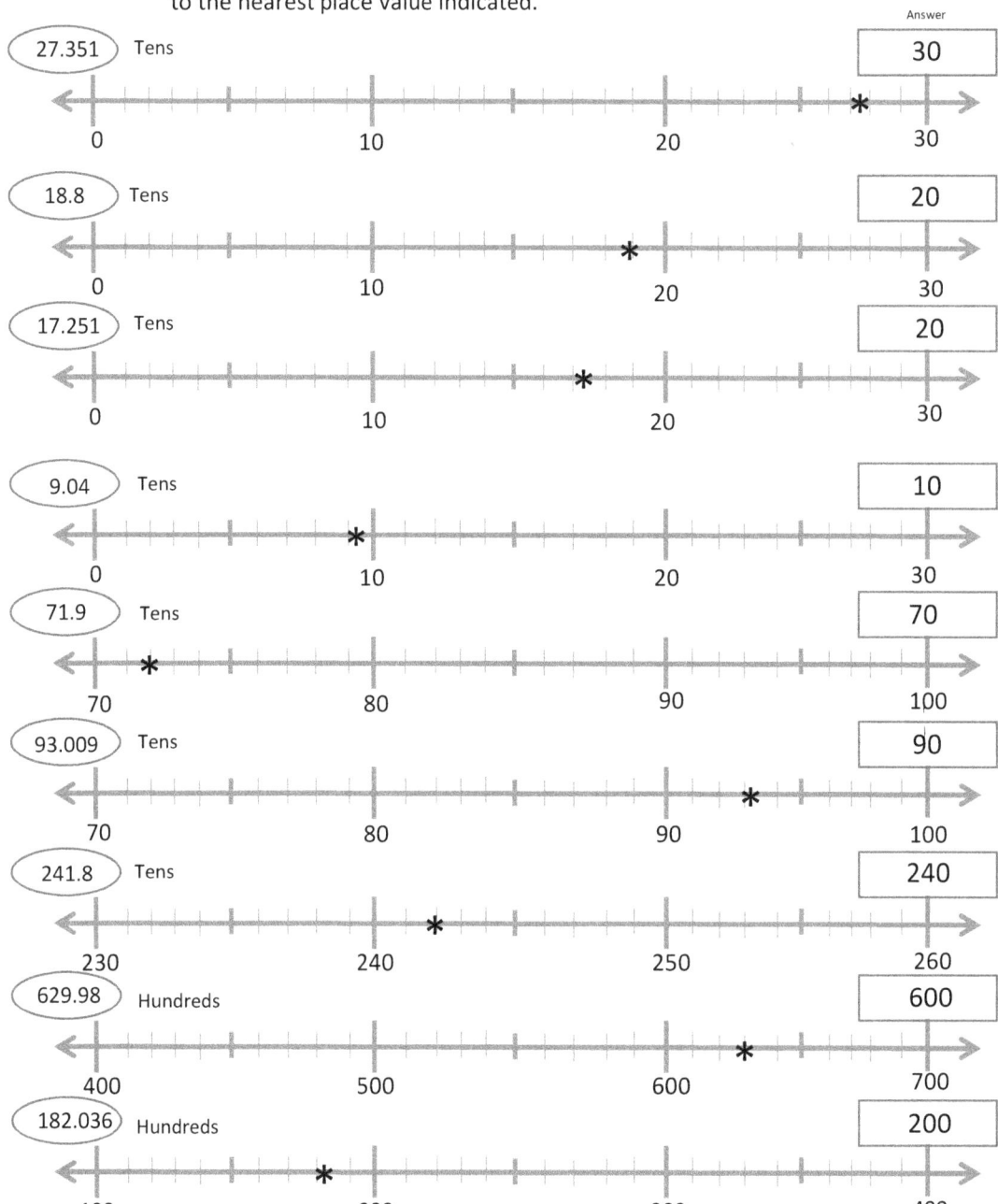

		Answer
(27.351)	Tens	30
18.8	Tens	20
(17.251)	Tens	20
9.04	Tens	10
71.9	Tens	70
(93.009)	Tens	90
241.8	Tens	240
(629.98)	Hundreds	600
(182.036)	Hundreds	200

Rounding Numbers (Estimation)

H 1

Name _____

Round the following numbers to the nearest ones place.

1) 963.62

2) 1,520.499

3) 63,123.801

Round the following numbers to the nearest whole number.

4) 100.832

5) 1,566.52498

6) 0.098

Round the following numbers to the nearest hundredths place.

7) 12.63289

8) 1,001.00531

9) 0.9963

Rounding Numbers (Estimation)

Name ___Key___

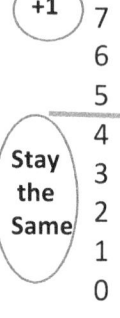

Round the following numbers
to the nearest ones place.

1. 963.62
   +1

   964

2. 1,520.499

   1,520

3. 63,123.801
   +1

   63,124

Round the following numbers
to the nearest whole number.

4. 100.832
   +1

   101

5. 1,566.52498
   +1

   1,567

6. 0.098

   0

Round the following numbers
to the nearest hundredths place.

7. 12.63289

   12.63

8. 1,001.00531
   +1

   1,001.01

9. 0.9963
   +1

   1.00

Rounding Numbers (Estimation)

Name _____

Ten Thousands	Thousands	Hundreds	Tens	Ones	And	Tenths	Hundredths
					.		

9
8
+1 7
6
5
4
Stay 3
the 2
Same 1
0

Round the following numbers
to the nearest tenths place.

(1) 963.62

(2) 1,520.499

(3) 63,123.801

_____ _____ _____

Round the following numbers
to the nearest hundredths.

(4) 100.832

(5) 1,566.52498

(6) 970.098

_____ _____ _____

Round the following numbers
to the nearest whole number.

(7) 12.63289

(8) 1,001.00531

(9) 0.9963

_____ _____ _____

Rounding Numbers (Estimation)

Name _____**Key**_____

Ten Thousands	Thousands	Hundreds	Tens	Ones	And	Tenths	Hundredths

Round the following numbers
to the nearest tenths place.

+1
5 ———
Stay the Same

9
8
7
6
5
4
3
2
1
0

(1) 963.62

936.6

(2) 1,520.499
     +1

1,520.5

(3) 63,123.801

63,123.8

Round the following numbers
to the nearest hundredths.

(4) 100.832

100.83

(5) 1,566.52498

1,566.52

(6) 970.098
     +1

970.10

Round the following numbers
to the nearest whole number.

(7) 12.63289
     +1

13

(8) 1,001.00531

1,001

(9) 0.9963
     +1

1

Estimate Numbers to the Nearest Hundredth

H 3

Name _____

9
8
(+1) 7
6
5
___
4
Stay
the 3
Same 2
1
0

1) 654.3827 _____

2) 23.9161 _____

3) 0.227111 _____

4) 999.99499 _____

5) 73.10709 _____

6) 362.63851 _____

7) 9.007014 _____

8) 45.318 _____

9) 9.4444 _____

10) 0.321 _____

11) 99.999 _____

12) 56,987,231.015 _____

Ten Thousands	Thousands	Hundreds	Tens	Ones	And	Tenths	Hundredths
					.		

Estimate Numbers to the Nearest Hundredth

Name _____ Key _____

9
8
⊙ +1   7
6
5
___
4
⬭ Stay the Same   3
2
1
0

1)   654.3827   654.38

2)   23.9161   23.92

3)   0.227111   0.23

4)   999.99499   999.99

5)   73.10709   73.11

6)   362.63851   362.64

7)   9.007014   9.01

8)   45.318   45.32

9)   9.4444   9.44

10) 0.321   0.32

11) 99.999   100

12) 56,987,231.015   56,987,231.02

Ten Thousands	Thousands	Hundreds	Tens	Ones	And	Tenths	Hundredths
					.		

Yellow Pencil Mathematics ™

Find the About, Almost, Approximate Answer

H 4

Name _____

1) 72.345 + 7.9214 =

2) 3.1999 + 754.889 =

3) 25.0987 + 5 =

4) 49.901 + 50.29 =

5) 7 + 3.018 =

6) 234.9 + 5 =

7) 300.634 + 99.7732 =

8) 2.0078 + 1.346 =

9) 100.482 − 99.63 =

10) 82 − 12.349 =

11) 327.9103 − 27.71 =

12) 27.81 − 17.6 =

13) 1000.276 − 999.096 =

14) 102.927 − 3.317 =

15) 56 − 20.4908 =

16) 3.63 − 2.7 =

Find the About, Almost, Approximate Answer

Name _____ **Key** _____

1. $72.345 + 7.9214 =$ 80
   $72 + 8$

2. $3.1999 + 754.889 =$ 758
   $3 + 755$

3. $25.0987 + 5 =$ 30
   $25 + 5$

4. $49.901 + 50.29 =$ 100
   $50 + 50$

5. $7 + 3.018 =$ 10
   $7 + 3$

6. $234.9 + 5 =$ 240
   $235 + 5$

7. $300.634 + 99.7732 =$ 401
   $301 + 100$

8. $2.0078 + 1.346 =$ 3
   $2 + 1$

9. $100.482 - 99.63 =$ 0
   $100 - 100$

10. $82 - 12.349 =$ 70
    $82 - 12$

11. $327.9103 - 27.71 =$ 300
    $328 - 28$

12. $27.81 - 17.6 =$ 10
    $28 - 18$

13. $1000.276 - 999.096 =$ 1
    $1000 - 999$

14. $102.927 - 3.317 =$ 100
    $103 - 3$

15. $56 - 20.4908 =$ 36
    $56 - 20$

16. $3.63 - 2.7 =$ 1
    $4 - 3$

Find the About, Almost, Approximate Answer

H 5

Name _____

1. 70.345 X 7.9214 =

2. 3.1999 X 759.889 =

3. 25.0987 X 5 =

4. 49.901 X 50.29 =

5. 7 X 3.018 =

6. 299.9 X 5 =

7. 300.234 X 99.7732 =

8. 2.0078 X 1.346 =

9. 100.482 X 99.63 =

10. 80 X 12.349 =

11. 329.9103 X 29.71 =

12. 27.81 X 19.6 =

13. 1000.276 X 999.096 =

14. 102.927 X 3.317 =

15. 56 X 20.4908 =

16. 3.63 X 2.7 =

Find the About, Almost, Approximate Answer

H 5

Name _____ Key _____

1) 70.345 X 7.9214 = 560
   70 X 8

2) 3.1999 X 759.889 = 2,280
   3 X 760

3) 25.0987 X 5 = 125
   25 X 5

4) 49.901 X 50.29 = 2,500
   50 X50

5) 7 X 3.018 = 21
   7 X 3

6) 299.9 X 5 = 1,500
   300 X 5

7) 300.234 X 99.7732 = 30,000
   300 X 100

8) 2.0078 X 1.346 = 2
   2 X 1

9) 100.482 X 99.63 = 10,000
   100 X 100

10) 80 X 12.349 = 960
    80 X 12

11) 329.9103 X 29.71 = 9,900
    330 X 30

12) 27.81 X 19.6 = 560
    28 X 20

13) 1000.276 X 999.096 =
    1000 X 1000 = 1,000,000

14) 102.927 X 3.317 = 309
    103 X 3

15) 56 X 20.4908 = 1,120
    56 X 20

16) 3.63 X 2.7 = 12
    4 X 3

Least Common Multiple   (LCM)

Name _____

			LCM
1)	8 6		
2)	4 10		
3)	6 4		
4)	9 6		
5)	12 8		
6)	6 9		

Least Common Multiple   (LCM)

Name ___KEY___

				LCM
1)	8	8, 16, (24)		24
	6	6, 12, 18, (24)		
2)	4	4, 8, 12, 16, (20)		20
	10	10, (20)		
3)	6	6, (12), 18		12
	4	4, 8, (12)		
4)	9	9, (18)		18
	6	6, 12, (18)		
5)	12	12, (24)		24
	8	8, 16, (24)		
6)	6	6, 12, (18)		18
	9	9, (18)		

Adding Fractions With Unlike Denominators

Name _____

1)	LCM $\dfrac{4}{3}$		
	$\dfrac{1}{4}$ + $\dfrac{2}{3}$		
2)	LCM $\dfrac{3}{9}$		
	$\dfrac{2}{3}$ + $\dfrac{1}{9}$		
3)	LCM $\dfrac{8}{3}$		
	$\dfrac{1}{8}$ + $\dfrac{2}{3}$		
4)	LCM		
	$\dfrac{3}{4}$ + $\dfrac{1}{6}$		
5)	LCM		
	$\dfrac{3}{8}$ + $\dfrac{3}{10}$		
6)	LCM		
	$\dfrac{1}{2}$ + $\dfrac{1}{4}$		

Adding Fractions With Unlike Denominators

1)	LCM $\dfrac{4}{3}$	4, 8, 12 3, 6, 9, 12	12
	$\dfrac{1^{\times 3}}{4_{\times 3}} + \dfrac{2^{\times 4}}{3_{\times 4}}$	$\dfrac{3}{12} + \dfrac{8}{12} = \dfrac{11}{12}$	$\dfrac{11}{12}$
2)	LCM $\dfrac{3}{9}$	3, 6, 9 9	9
	$\dfrac{2^{\times 3}}{3_{\times 3}} + \dfrac{1^{\times 1}}{9_{\times 1}}$	$\dfrac{6}{9} + \dfrac{1}{9} = \dfrac{7}{9}$	$\dfrac{7}{9}$
3)	LCM $\dfrac{8}{3}$	8, 16, 24 3, 6, 9, 12, 15, 18, 21, 24	24
	$\dfrac{1^{\times 3}}{8_{\times 3}} + \dfrac{2^{\times 8}}{3_{\times 8}}$	$\dfrac{3}{24} + \dfrac{16}{24} = \dfrac{19}{24}$	$\dfrac{19}{24}$
4)	LCM $\dfrac{4}{6}$	4, 8, 12 6, 12	12
	$\dfrac{3^{\times 3}}{4_{\times 3}} + \dfrac{1^{\times 2}}{6_{\times 2}}$	$\dfrac{9}{12} + \dfrac{2}{12} = \dfrac{11}{12}$	$\dfrac{11}{12}$
5)	LCM $\dfrac{8}{10}$	8, 16, 24, 32, 40 10, 20, 30, 40	40
	$\dfrac{3^{\times 5}}{8_{\times 5}} + \dfrac{3^{\times 4}}{10_{\times 4}}$	$\dfrac{15}{40} + \dfrac{12}{40} = \dfrac{27}{40}$	$\dfrac{27}{40}$
6)	LCM $\dfrac{2}{4}$	2, 4 4	4
	$\dfrac{1^{\times 2}}{2_{\times 2}} + \dfrac{1^{\times 1}}{4_{\times 1}}$	$\dfrac{2}{4} + \dfrac{1}{4} = \dfrac{3}{4}$	$\dfrac{3}{4}$

Yellow Pencil Mathematics™

Subtracting Fractions With Unlike Denominators

Name _____

LCM $\dfrac{8}{6}$		
1) $\dfrac{7}{8} - \dfrac{5}{6}$		
LCM $\dfrac{6}{12}$		
2) $\dfrac{5}{6} - \dfrac{7}{12}$		
LCM $\dfrac{4}{2}$		
3) $\dfrac{3}{4} - \dfrac{1}{2}$		
LCM		
4) $\dfrac{11}{12} - \dfrac{1}{3}$		
LCM		
5) $\dfrac{3}{4} - \dfrac{3}{5}$		
LCM		
6) $\dfrac{8}{9} - \dfrac{1}{6}$		

97

Subtracting Fractions With Unlike Denominators

I 3

Name _____**Key**_____

1)	LCM $\dfrac{8}{6}$	8, 16, 24 6, 12, 18, 24	24
	$\dfrac{7^{\times 3}}{8^{\times 3}} - \dfrac{5^{\times 4}}{6^{\times 4}}$	$\dfrac{21}{24} - \dfrac{20}{24} = \dfrac{1}{24}$	$\dfrac{1}{24}$
2)	LCM $\dfrac{6}{12}$	6, 12, 18, 24 12, 24	12
	$\dfrac{5^{\times 2}}{6^{\times 2}} - \dfrac{7^{\times 1}}{12^{\times 1}}$	$\dfrac{10}{12} - \dfrac{7}{12} = \dfrac{3}{12}$	$\dfrac{3}{12}$
3)	LCM $\dfrac{4}{2}$	4 2, 4	4
	$\dfrac{3^{\times 1}}{4^{\times 1}} - \dfrac{1^{\times 2}}{2^{\times 2}}$	$\dfrac{3}{4} - \dfrac{2}{4} = \dfrac{1}{4}$	$\dfrac{1}{4}$
4)	LCM $\dfrac{12}{3}$	12 3, 6, 9, 12	12
	$\dfrac{11^{\times 1}}{12^{\times 1}} - \dfrac{1^{\times 4}}{3^{\times 4}}$	$\dfrac{11}{12} - \dfrac{4}{12} = \dfrac{7}{12}$	$\dfrac{7}{12}$
5)	LCM $\dfrac{4}{5}$	4, 8, 12, 16, 20 5, 10, 15, 20	20
	$\dfrac{3^{\times 5}}{4^{\times 5}} - \dfrac{3^{\times 4}}{5^{\times 4}}$	$\dfrac{15}{20} - \dfrac{12}{20} = \dfrac{3}{20}$	$\dfrac{3}{20}$
6)	LCM $\dfrac{9}{6}$	9, 18 6, 12, 18	18
	$\dfrac{8^{\times 2}}{9^{\times 2}} - \dfrac{1^{\times 3}}{6^{\times 3}}$	$\dfrac{16}{18} - \dfrac{3}{18} = \dfrac{13}{18}$	$\dfrac{13}{18}$

Yellow Pencil Mathematics ™

Adding and Subtracting Fractions With Unlike Denominators

I 4

Name _____

1) $\dfrac{2}{5} + \dfrac{1}{4}$		
2) $\dfrac{3}{4} + \dfrac{1}{6}$		
3) $\dfrac{1}{3} + \dfrac{4}{9}$		
4) $\dfrac{5}{6} - \dfrac{1}{3}$		
5) $\dfrac{4}{9} - \dfrac{1}{6}$		
6) $\dfrac{4}{5} - \dfrac{2}{3}$		

Adding and Subtracting Fractions With Unlike Denominators

I 4 Name ___KEY___

1)	LCM $\dfrac{5}{4}$	5, 10, 15, 20 4, 8, 12, 16, 20		20
	$\dfrac{2^{\times 4}}{5^{\times 4}} + \dfrac{1^{\times 5}}{4^{\times 5}}$	$\dfrac{8}{20} + \dfrac{5}{20} = \dfrac{13}{20}$		$\dfrac{13}{20}$
2)	LCM $\dfrac{4}{6}$	4, 8, 12 6, 12		12
	$\dfrac{3^{\times 3}}{4^{\times 3}} + \dfrac{1^{\times 2}}{6^{\times 2}}$	$\dfrac{9}{12} + \dfrac{2}{12} = \dfrac{11}{12}$		$\dfrac{11}{12}$
3)	LCM $\dfrac{3}{9}$	3, 6, 9 9		9
	$\dfrac{1^{\times 3}}{3^{\times 3}} + \dfrac{4^{\times 1}}{9^{\times 1}}$	$\dfrac{3}{9} + \dfrac{4}{9} = \dfrac{7}{9}$		$\dfrac{7}{9}$
4)	LCM $\dfrac{6}{3}$	6 3, 6		6
	$\dfrac{5^{\times 1}}{6^{\times 1}} - \dfrac{1^{\times 2}}{3^{\times 2}}$	$\dfrac{5}{6} - \dfrac{2}{6} = \dfrac{3}{6}$		$\dfrac{3}{6}$
5)	LCM $\dfrac{9}{6}$	9, 18 6, 12, 18		18
	$\dfrac{4^{\times 2}}{9^{\times 2}} - \dfrac{1^{\times 3}}{6^{\times 3}}$	$\dfrac{8}{18} - \dfrac{3}{18} = \dfrac{5}{18}$		$\dfrac{5}{18}$
6)	LCM $\dfrac{5}{3}$	5, 10, 15 3, 6, 9, 12, 15		15
	$\dfrac{4^{\times 3}}{5^{\times 3}} - \dfrac{2^{\times 5}}{3^{\times 5}}$	$\dfrac{12}{15} - \dfrac{10}{15} = \dfrac{2}{15}$		$\dfrac{2}{15}$

64 Addition Facts

7 + 9	9 + 9	8 + 3	6 + 7	6 + 2	6 + 5	8 + 9	1 + 9
9 + 4	7 + 3	6 + 6	2 + 4	5 + 2	7 + 7	5 + 6	9 + 7
3 + 2	8 + 4	4 + 7	9 + 8	4 + 6	4 + 5	7 + 5	2 + 2
6 + 3	9 + 3	8 + 8	9 + 6	7 + 4	8 + 5	3 + 5	9 + 6
9 + 2	2 + 8	9 + 5	2 + 6	3 + 8	6 + 9	5 + 8	3 + 4
3 + 7	4 + 8	1 + 7	4 + 3	1 + 6	2 + 5	4 + 4	7 + 6
7 + 8	4 + 9	2 + 9	6 + 4	7 + 2	8 + 6	5 + 4	8 + 2
5 + 3	6 + 8	3 + 9	1 + 8	5 + 7	8 + 7	5 + 9	2 + 7

Yellow Pencil Mathematics ™

64 Subtraction Facts

0 1 2 3 4 5 6 7 8 9 10 11 12 13 14 15 16 17 18 19 20

16 - 9	18 - 9	11 - 3	13 - 7	8 - 2	11 - 5	17 - 9	10 - 9
13 - 4	10 - 3	12 - 6	6 - 4	7 - 2	14 - 7	11 - 6	16 - 7
5 - 2	12 - 4	11 - 7	17 - 8	10 - 6	9 - 5	12 - 5	4 - 2
9 - 3	12 - 3	16 - 8	15 - 6	11 - 4	13 - 5	8 - 5	9 - 6
11 - 2	10 - 8	14 - 5	8 - 6	11 - 8	15 - 9	13 - 8	7 - 4
10 - 7	12 - 8	8 - 7	7 - 3	7 - 6	7 - 5	8 - 4	13 - 6
15 - 8	13 - 9	11 - 9	10 - 4	9 - 2	14 - 6	9 - 4	10 - 2
8 - 3	14 - 8	12 - 9	9 - 8	12 - 7	15 - 7	14 - 9	9 - 7

36 Facts in Order 2s Through 9s

Time_____   Name _____

| 2 X 2 | 2 X 3 | 2 X 4 | 2 X 5 | 2 X 6 | 2 X 7 | 2 X 8 | 2 X 9 |

| 3 X 3 | 3 X 4 | 3 X 5 | 3 X 6 | 3 X 7 | 3 X 8 | 3 X 9 |

| 4 X 4 | 4 X 5 | 4 X 6 | 4 X 7 | 4 X 8 | 4 X 9 |

| 5 X 5 | 5 X 6 | 5 X 7 | 5 X 8 | 5 X 9 |

| 6 X 6 | 6 X 7 | 6 X 8 | 6 X 9 |

| 7 X 7 | 7 X 8 | 7 X 9 |

| 8 X 8 | 8 X 9 |

| 9 X 9 |

36 Mixed Multiplication Facts 2s Through 9s

4 X 7	6 X 8	5 X 9	4 X 4	2 X 3	2 X 8
3 X 7	5 X 7	8 X 9	2 X 4	5 X 8	3 X 5
6 X 7	8 X 8	2 X 6	4 X 9	5 X 6	3 X 3
9 X 9	6 X 9	2 X 2	3 X 9	7 X 8	6 X 6
4 X 6	3 X 6	7 X 7	3 X 4	5 X 5	4 X 8
7 X 9	4 X 5	2 X 5	3 X 8	2 X 7	2 X 9

Yellow Pencil Mathematics ™

64 Mixed Multiplication Facts 2s Through 9s

Time_____   Name _____

3 X 7	7 X 4	2 X 3	3 X 5	9 X 8	7 X 2	8 X 3	4 X 4
8 X 8	6 X 3	2 X 9	6 X 5	5 X 9	3 X 4	8 X 9	4 X 8
3 X 6	4 X 9	6 X 7	2 X 5	5 X 6	4 X 3	6 X 6	8 X 7
4 X 2	4 X 6	8 X 5	9 X 4	3 X 8	9 X 5	6 X 9	7 X 5
8 X 4	5 X 7	2 X 6	8 X 6	9 X 9	7 X 6	6 X 4	5 X 8
7 X 8	5 X 2	9 X 3	5 X 4	7 X 9	4 X 5	6 X 8	7 X 3
7 X 2	3 X 3	9 X 2	5 X 5	2 X 4	7 X 7	3 X 2	9 X 7
8 X 2	3 X 9	4 X 7	9 X 6	5 X 3	6 X 2	2 X 2	2 X 8

Yellow Pencil Mathematics ™

Multiple Practice

Insert the first ten multiples of each number.

1									
2									
3									
4									
5									
6									
7									
8									
9									
10									

Multiplication Chart

X	1	2	3	4	5	6	7	8	9	10
1	1	2	3	4	5	6	7	8	9	10
2	2	4	6	8	10	12	14	16	18	20
3	3	6	9	12	15	18	21	24	27	30
4	4	8	12	16	20	24	28	32	36	40
5	5	10	15	20	25	30	35	40	45	50
6	6	12	18	24	30	36	42	48	54	60
7	7	14	21	28	35	42	49	56	63	70
8	8	16	24	32	40	48	56	64	72	80
9	9	18	27	36	45	54	63	72	81	90
10	10	20	30	40	50	60	70	80	90	100

Division Starter

Example: What times 3 equals 21?

X 3 21	X 7 28	X 2 6	X 3 15	X 9 72	X 7 14	X 8 24	X 4 16
X 8 64	X 6 18	X 2 18	X 6 30	X 5 45	X 3 12	X 8 72	X 4 32
X 3 18	X 4 36	X 6 42	X 2 10	X 5 30	X 4 12	X 6 36	X 8 56
X 4 8	X 4 24	X 8 40	X 9 36	X 3 24	X 9 45	X 6 54	X 7 35
X 8 32	X 5 35	X 2 12	X 8 48	X 9 81	X 7 42	X 6 24	X 5 40
X 7 56	X 5 10	X 9 27	X 5 20	X 7 63	X 4 20	X 6 48	X 7 21
X 7 14	X 3 9	X 9 18	X 5 25	X 2 8	X 7 49	X 3 6	X 9 63
X 8 16	X 3 27	X 4 28	X 9 54	X 5 15	X 6 12	X 2 4	X 2 16

Division Practice

$\dfrac{21}{3}$	$\dfrac{28}{7}$	$\dfrac{6}{2}$	$\dfrac{15}{3}$	$\dfrac{72}{9}$	$\dfrac{14}{7}$	$\dfrac{24}{8}$	$\dfrac{16}{4}$
$\dfrac{64}{8}$	$\dfrac{18}{6}$	$\dfrac{18}{2}$	$\dfrac{30}{6}$	$\dfrac{45}{5}$	$\dfrac{12}{3}$	$\dfrac{72}{8}$	$\dfrac{32}{4}$
$\dfrac{18}{3}$	$\dfrac{36}{4}$	$\dfrac{42}{6}$	$\dfrac{10}{2}$	$\dfrac{30}{5}$	$\dfrac{12}{4}$	$\dfrac{36}{6}$	$\dfrac{56}{8}$
$\dfrac{8}{4}$	$\dfrac{24}{4}$	$\dfrac{40}{8}$	$\dfrac{36}{9}$	$\dfrac{24}{3}$	$\dfrac{45}{9}$	$\dfrac{54}{6}$	$\dfrac{35}{7}$
$\dfrac{32}{8}$	$\dfrac{35}{5}$	$\dfrac{12}{2}$	$\dfrac{48}{8}$	$\dfrac{81}{9}$	$\dfrac{42}{7}$	$\dfrac{24}{6}$	$\dfrac{40}{5}$
$\dfrac{56}{7}$	$\dfrac{10}{5}$	$\dfrac{27}{9}$	$\dfrac{20}{5}$	$\dfrac{63}{7}$	$\dfrac{20}{4}$	$\dfrac{48}{6}$	$\dfrac{21}{7}$
$\dfrac{14}{7}$	$\dfrac{9}{3}$	$\dfrac{18}{9}$	$\dfrac{25}{5}$	$\dfrac{8}{2}$	$\dfrac{49}{7}$	$\dfrac{6}{3}$	$\dfrac{63}{9}$
$\dfrac{16}{8}$	$\dfrac{27}{3}$	$\dfrac{28}{4}$	$\dfrac{54}{9}$	$\dfrac{15}{5}$	$\dfrac{12}{6}$	$\dfrac{4}{2}$	$\dfrac{16}{2}$

Math Intervention Report

Student _____ Year _____ Teacher _____

State Test Score _____

Skills Screening Score: September _____ February _____ May _____

Multiplication Fact Proficiency as determined by being able to correctly complete 64 facts (all 2s through 9s) in less than five minutes without accommodations:

September _____   February _____   May _____

Mathematical Deficit: The student has difficulty solving basic Addition Facts and solving three-digit by three-digit addition problems.

Intervention: Have student practice the Basic Addition Practice paper using a number line. Have student practice putting basic three-digit by three-digit problems in a Place Value chart and then solve for the answer.

Mathematical Deficit: The student has difficulty solving basic Subtraction Facts and solving three-digit by three-digit subtraction problems.

Intervention: Have student practice the Basic Subtraction paper using a number line. Have student practice putting basic three-digit by three-digit problems in a Place Value chart and then solve for the answer.

Mathematical Deficit: The student has difficulty solving basic Multiplication Facts and solving three-digit by two-digit multiplication problems.

Intervention: Have student practice the 36 Basic Multiplication Facts in Order 2s through 9s paper, the 36 Basic Multiplication Facts Mixed 2s through 9s paper, and the 64 Basic Multiplication Mixed paper. Have students take a timed test on the 36 Basic Multiplication Facts in Order 2s through 9s paper bi-weekly and take a timed test on the 64 Basic Multiplication Mixed paper monthly. Have students practice multiplying three-digit by two-digit multiplication problems in grid paper.

Mathematical Deficit: The student has difficulty solving basic Division Facts and solving problems with a three-digit dividend and a single-digit divisor.

Intervention: Have student practice the Division Starter paper—backward multiplication—and the 64 Basic Division Facts paper. Teach students to divide using long division by making a, MULTIPLE, list of the divisor.

Mathematical Deficit: The student has difficulty with Place Value

Intervention: Have student practice putting numbers in a Place Value Chart and then naming the place value of different digits. Place Value is also taught with the Addition and Subtraction papers mentioned above and the Expanded Form, Expanded Notation, and Rounding papers mentioned below.

Mathematical Deficit: The student has difficulty putting numbers into Expanded Form and Expanded Notation.

Intervention: Have student decompose numbers into a Place Value Chart and then practice putting numbers into Expanded Form and Expanded Notation. Have students practice Expanded Form before practicing Expanded Notation.

Mathematical Deficit: The student has difficulty manipulating numbers on Number Lines.

Intervention: Have student practice manipulating numbers on number lines.

Mathematical Deficit: The student has difficulty rounding numbers.

Intervention: Have student practice placing numbers into a place value chart and then rounding for a given place value. Teach student to round using a number line.

Mathematical Deficit: The student has difficulty adding and subtracting simple fractions with unlike denominators.

Intervention: Have student practice completing the Multiples Chart. Use the knowledge learned by practicing the Multiples Chart to list Least Common Multiples, LCM, in the Least Common Multiples lesson. Use LCM to find the Greatest Comon Denominator, GCD, of fractions with unlike denominators. Teach students to convert to fractions with like denominators and add or subtract numerators.

Name _____   Class _____

1. 5,317 + 693 =

A. 6,115
B. 9,925
C. 6,235
D. 6,010

2. 8,956 + 7,435 =

F. 15,989
G. 16,387
H. 16,391
J. 16,487

3. 6,928 + 875 =

A. 8,734
B. 7,803
C. 7,892
D. 7,832

4. 8,564 − 586 =

F. 7,978
G. 8,797
H. 7,869
J. 8,025

5. 6,143 − 98 =

A. 6,027
B. 6,045
C. 5,904
D. 6,524

6. 3,001 - 164 =

    F. 2,837
    G. 2,942
    H. 2,831
    J. 2,824

7. 934 X 59 =

    A. 55,421
    B. 55,106
    C. 55,614
    D. 55,425

8. 891 X 36 =

    F. 31,134
    G. 32,147
    H. 32,076
    J. 32,206

9. 354 X 27 =

    A. 9,305
    B. 9,552
    C. 10,214
    D. 9,558

10. 621 divided by 3 =

    F. 198
    G. 214
    H. 200
    J. 207

11.   1,788 divided by 6 =

A. 298
B. 304
C. 292
D. 311

12.   2,884 divided by 4 =

F. 729
G. 721
H. 718
J. 709

13.   In the number 9,635 where does the decimal go?

A. The decimal goes in between the three and five.
B. The decimal goes to the left of the nine.
C. The decimal goes to the right of the five.
D. The decimal goes in between the six and three.

14.   In the number 1,024.56 the 0 is in what place value?

F. Hundredths
G. Tenths
H. Hundreds
J. Tens

15.   In the number 1,024.56 the 6 is in what place value?

A. Hundreds
B. Tenths
C. Hundredths
D. Tens

16.   How is one thousand, one hundred ninety-nine written in expanded form?

  F. 100 + 90 + 9
  G. 1,000 + 90 + 9
  H. 1,000 + 100 + 9
   J. 1,000 + 100 + 90 + 9

17.   How is the number 963.05 written in Expanded Notation?

  A. (9 X 100) + (3 X 1) + (0 X 0.1) + (5 X 0.01)
  B. (9 X 100) + (6 X 10) + (3 X 1) + (5 X 0.001)
  C. (9 X 100) + (6 X 10) + (3 X 1) + (5 X 0.01)
  D. (9 X 1000) + (6 X 10) + (3 X 1) + (5 X 0.01)

18.   3,000 + 20 + 0.4 + 0.07 =

  F. 3,200.47
  G. 3,020.47
  H. 3002.47
   J. 3020.04

19. Which number best represents the point R on the number line?

0   100

        A. 90
        B. 87
        C. 92
        D. 95

20. Which number best represents the point R on the number line?

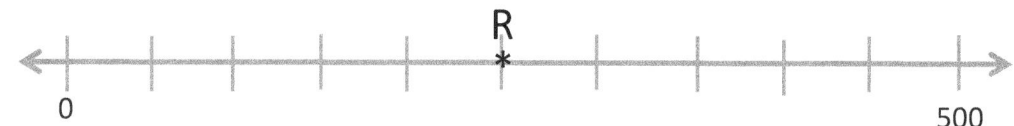

    F. 550
    G. 500
    H. 300
    J. 250

21. Which number best represents the point R on the number line?

    A. 875
    B. 925
    C. 775
    D. 750

22. Round 8,271.29 to the tenths place.

    F. 8,272
    G. 8,271.3
    H. 8,271
    J. 8,270.2

23. Round 8,271.29 to the hundreds place.

    A. 8,270
    B. 8,300
    C. 8,271.3
    D. 8,271.2

24. Round 8,271.29 to the ones place.

    F. 8,300
    G. 8,271.3
    H. 8,271
    J. 8,270

---

25. $$\frac{1}{4} + \frac{3}{8} =$$

A. $\frac{4}{12}$ B. $\frac{21}{32}$ C. $\frac{5}{8}$ D. $\frac{1}{2}$

---

26. $$\frac{7}{9} - \frac{1}{3} =$$

F. $\frac{4}{9}$ G. $\frac{6}{6}$ H. $\frac{5}{27}$ J. $\frac{10}{9}$

---

27. $$\frac{1}{6} + \frac{1}{8} =$$

A. $\frac{11}{48}$ B. $\frac{2}{14}$ C. $\frac{3}{8}$ D. $\frac{7}{24}$

Fifth Grade Response to Intervention Skills Test
Middle Screening

Name _____   Class _____

1.   627 + 21,593 =

A. 22,210
B. 22,301
C. 22,190
D. 22,220

2.   761 + 9,429 =

F. 10,111
G. 10,190
H. 10,238
J. 10,130

3.   9,989 + 1,128

A. 11,117
B. 11,103
C. 11,189
D. 10,359

4.   9,026 – 8139 =

F. 1,021
G. 892
H. 854
J. 887

5.   6,143 – 982

A. 5,561
B. 5,184
C. 5,161
D. 5,267

6.   5,000 - 111 =

F. 4,476
G. 4,895
H. 4,789
J. 4,889

7.   934 X 22 =

A. 20,548
B. 22,348
C. 20,624
D. 21,348

8.   567 X 89 =

F. 50,463
G. 50.589
H. 49,563
J. 51,692

9.   376 X 72 =

A. 27,172
B. 27,072
C. 28,034
D. 27,432

10.   938 divided by 7 =

F. 324
G. 128
H. 206
J. 134

11.   4,935 divided by 7 =

A. 722
B. 705
C. 608
D. 675

12.   1,398 divided by 6 =

F. 203
G. 245
H. 233
J. 235

13.   In the number 0372 where does the decimal go?

A. The decimal goes in between the zero and three.
B. The decimal goes to the left of the zero.
C. The decimal goes to the right of the two.
D. The decimal goes in between the three and seven.

14.   In the number 1,024.56 the 5 is in what place value?

F. Hundredths
G. Tenths
H. Tens
J. Ones

15.   In the number 1,024.56 the 2 is in what place value?

A. Hundreds
B. Tenths
C. Hundredths
D. Tens

16.   1,000 + 100 + 90 + 9 =

   F. One thousand, one hundred nine
   G. Two thousand, ninety-nine
   H. One thousand, one hundred ninety-nine
   J. One thousand, nine hundred ninety

17.   How is the number 9,321.25 written in Expanded Form?

   A. 9,000 + 300 + 20 + 0.2 + 0.05
   B. 9,000 + 300 + 20 + 1 + 0.2 + 0.05
   C. 9,000 + 300 + 20 + 1 + 0.25
   D. 9,000 + 300 + 20 + 1 + 0.02 + 0.005

18.   How is the number one hundred six and seven hundredths written in
      Expanded Notation?

   F. (1 X 100) + (6 X 10) + (7 X 0.01)
   G. (1 X 100) + (6 X 10) + (7 X 0.001)
   H. (1 X 100) + (6 X 1) + (7 X 0.001)
   J. (1 X 100) + (6 X 1) + (7 X 0.01)

19.   Which number best represents the point R on the number line?

0   100

      A. 60
      B. 75
      C. 65
      D. 70

Fifth Grade Response to Intervention Skills Test
Middle Screening

20. Which number best represents the point R on the number line?

0   500

      F. 225
      G. 650
      H. 325
      J. 725

---

21. Which number best represents the point R on the number line?

0   1000

      A. 875
      B. 925
      C. 775
      D. 750

---

22. Round 8,271.29 to the ones place.

      F. 8,272
      G. 8,271.3
      H. 8,271
      J. 8,270.2

---

23. Round 8,271.29 to the tenths place.

      A. 8,272.0
      B. 8,271
      C. 8,271.3
      D. 8,271.2

24. Round 346 to the ones place.

  F. 300
  G. 347
  H. 340
  J. 346

25.
$$\frac{1}{6} + \frac{4}{9} =$$

A. $\dfrac{5}{54}$ B. $\dfrac{3}{4}$ C. $\dfrac{5}{15}$ D. $\dfrac{11}{18}$

26.
$$\frac{5}{8} - \frac{5}{12} =$$

F. $\dfrac{5}{24}$ G. $\dfrac{1}{2}$ H. $\dfrac{10}{96}$ J. $\dfrac{10}{24}$

27.
$$\frac{3}{10} + \frac{3}{8} =$$

A. $\dfrac{1}{3}$ B. $\dfrac{27}{40}$ C. $\dfrac{6}{18}$ D. $\dfrac{17}{80}$

Fifth Grade Response to Intervention Skills Test
End Screening

Name _____   Class _____

1.   8,372 + 2,849 =

A. 10,962
B. 11,221
C. 12,324
D. 11,321

2.   7,892 + 3,149 =

F. 11,041
G. 10,634
H. 11,782
J. 11,541

3.   8,351 + 1,429 =

A. 9,680
B. 9,730
C. 9,780
D. 9,124

4.   3,137 − 216 =

F. 2,921
G. 2,621
H. 2,937
J. 3,041

5.   6,422 − 5,288

A. 1,424
B. 1,124
C. 1,034
D. 1,134

6. 1,067 - 836 =

F. 132
G. 247
H. 231
J. 241

7. 327 X 82 =

A. 26,723
B. 35,814
C. 26,814
D. 27,063

8. 463 X 34 =

F. 17,642
G. 15,742
H. 21,244
J. 16,142

9. 564 X 83 =

A. 64,812
B. 45,894
C. 46,314
D. 46,812

10. 932 divided by 4 =

F. 233
G. 221
H. 323
J. 307

11.   3,510 divided by 5 =

    A. 834
    B. 702
    C. 712
    D. 804

---

12.   4,344 divided by 8 =

    F. 543
    G. 521
    H. 673
    J. 641

---

13.   In the number 304 where does the decimal go?

    A. The decimal goes in between the zero and four.
    B. The decimal goes to the left of the zero.
    C. The decimal goes to the right of the four.
    D. The decimal goes to the left of the three.

---

14.   In the number 86,531.745 the 4 is in what place value?

    F. Hundreds
    G. Tenths
    H. Tens
    J. Hundredths

---

15.   In the number 806.371 the 0 is in what place value?

    A. Hundreds
    B. Tenths
    C. Hundredths
    D. Tens

16.   4,000 + 70 + 2 + 0.04 =

   F. Four thousand seven hundred twenty and four hundredths
   G. Four thousand seventy-two and four hundredths
   H. Four thousand seventy-two and four tenths
   J. Four thousand twenty seven and twenty-four hundredths

17.   How is the number 10,304.6 written in Expanded Form?

   A. 10,000 + 30 + 4 + 0.06
   B. 10,000 + 300 + 4 + 0.06
   C. 10,000 + 300 + 4 + 0.6
   D. 10,000 + 200 + 4 + 0.6

18.   How is the number three thousand, two hundred forty and seven tenths
      written in Expanded Notation?

   F. (3 X 1000) + (2 X 100) + (4 X 10) + (7 X 0.1)
   G. (3 X 1000) + (2 X 10) + (4 X 1) + (7 X 0.01)
   H. (3 X 1000) + (2 X 100) + (4 X 10) + (7 X 0.01)
   J. (3 X 100) + (2 X 10) + (4 X 1) + (7 X 0.1)

19. Which number best represents the point R on the number line?

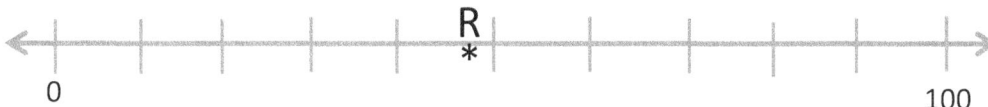

0   100

      A. 50
      B. 42
      C. 47
      D. 53

20. Which number best represents the point R on the number line?

F. 225
G. 450
H. 325
J. 350

21. Which number best represents the point R on the number line?

A. 200
B. 100
C. 150
D. 250

22. Round 324.74 to the tenths place.

F. 325
G. 324.7
H. 320
J. 324.8

23. Round 640.81 to the ones place.

A. 641
B. 651.8
C. 639
D. 640

24. Round 27,934.63 to the thousands place.

    F. 30,000
    G. 28,984
    H. 27,900
    J. 28,000

---

25. $\dfrac{5}{8} + \dfrac{1}{6} =$

A. $\dfrac{19}{24}$ B. $\dfrac{6}{14}$ C. $\dfrac{35}{48}$ D. $\dfrac{2}{7}$

---

26. $\dfrac{5}{6} - \dfrac{2}{9} =$

F. $\dfrac{3}{18}$ G. $\dfrac{11}{18}$ H. $\dfrac{39}{54}$ J. $\dfrac{1}{6}$

---

27. $\dfrac{1}{4} + \dfrac{3}{10} =$

A. $\dfrac{20}{40}$ B. $\dfrac{1}{2}$ C. $\dfrac{11}{20}$ D. $\dfrac{4}{14}$

Beginning, Middle, and End Screening Test Keys

Beginning		Middle		End	
1	D	1	D	1	B
2	H	2	G	2	F
3	B	3	A	3	C
4	F	4	J	4	F
5	B	5	C	5	D
6	F	6	J	6	H
7	B	7	A	7	C
8	H	8	F	8	G
9	D	9	B	9	D
10	J	10	J	10	F
11	A	11	B	11	B
12	G	12	H	12	F
13	C	13	C	13	C
14	H	14	G	14	J
15	C	15	D	15	D
16	J	16	H	16	G
17	C	17	B	17	C
18	G	18	J	18	F
19	C	19	C	19	C
20	J	20	H	20	F
21	A	21	A	21	C
22	G	22	H	22	G
23	B	23	C	23	A
24	H	24	J	24	J
25	C	25	D	25	A
26	F	26	F	26	G
27	D	27	B	27	C

www.ingramcontent.com/pod-product-compliance
Lightning Source LLC
Chambersburg PA
CBHW081238180526
45171CB00005B/463